T0189719

Research in Chemistry Education

Liliana Mammino • Jan Apotheker
Editors

Research in Chemistry Education

Editors
Liliana Mammino
School of Mathematical and Natural Sciences
University of Venda
Thohoyandou, South Africa

Jan Apotheker
Faculty of Science and Engineering
University of Groningen
Groningen, Groningen, The Netherlands

ISBN 978-3-030-59884-6 ISBN 978-3-030-59882-2 (eBook)
https://doi.org/10.1007/978-3-030-59882-2

This Springer imprint is published by the registered company Springer Nature Switzerland AG
The registered company address is: Gewerbestrasse 11, 6330 Cham, Switzerland

Preface

Chemistry is the science of substances. In everyday life, we use a variety of substances and materials for diverse purposes and with diverse effects; chemistry is at the basis of their production. Moreover, chemistry is the science that plays major roles in the efforts to address a number of crucial problems, from human health to sustainability. Chemistry research plays essential roles in the development of new medicines for the treatment of diseases, including emerging ones. It has crucial roles to play for the achievement of the United Nations Sustainable Development Goals. And it is at the bases of the search for improved substances and materials for everyday usage.

For chemistry research to be able to meet the demands from society, each country needs to have an adequate number of adequately prepared chemists; this, in turn, requires that learners are attracted to chemistry since their first encounter with it, so that a sufficient number of young people feel motivated to choose chemistry for their future profession. Furthermore, pursuing health and sustainability requires that citizens have adequate chemistry literacy, to be aware of the best and safest ways of handling substances and materials. The realization of all these objectives depends on the quality and effectiveness of chemistry education. Thus, the search for ways to continuously improve the quality and effectiveness of chemistry education plays fundamental roles for human well-being, for development, and for making development sustainable. The *African Conferences on Research in Chemistry Education* (ACRICE) have this aim for the African continent. They started in 2013, on the model of the longer established ECRICE (*European Conferences on Research in Chemistry Education*).

Chemistry as a natural science is context-independent, but chemistry education needs to take into account the 'real learners' and, therefore, it needs to take into account all the contextual features that might influence the quality and extent of learning. Continental-based conferences are broad enough to include a high variety of contexts, open enough to make opportunities for contributions from other continents, and focused enough to enable attentive consideration of local realities in the given continent and exchanges among educators facing similar challenges.

ACRICE-2 (recognized as an IUPAC conference) was held at the University of Venda (Thohoyandou, South Africa) from 22 to 27 November 2015. It gathered participants from several countries and continents (Africa, Asia, Europe, North and South America). These proceedings offer a selection of the papers that were presented at the conference, encompassing a broad range of themes. Chapter 1 has a general character, analysing the challenges and opportunities for Chemistry education in Africa, in the current globalization trend, with particular focus on the educational possibilities offered by modern technologies. Chapters 2 and 3 present microscale laboratory practices and their education and sustainability advantages. Chapter 4 outlines the application of systemic approaches to the teaching of organic chemistry. Chapters 5, 6, 7, 8, and 9 present diverse contextual experiences: a pre-course verification of students' background preparedness for an inorganic chemistry university course; the challenges related to the professional training of intermediate-level chemical analysts; the exploration of new assessment options; a verification of secondary school pupils' understanding of basic concepts on the microscopic world of atoms and molecules; the introduction of information about modern industrial research and production into secondary schools. Chapter 10 outlines possible pathways for the integration of chemistry education research into the normal teaching activities, with an action-research-type framework. Chapter 11 outlines the challenges of teaching chemistry students a different science (physics), whose background knowledge is necessary for the training of chemists. Chapter 12 has a different nature, as it was written by students; it is meant as documentation of how students can be brought to share in the motivations for the selection of specific educational approaches, and it narrates their active engagement in building their knowledge about green chemistry principles. Chapter 13 has a sort of overview roles on key challenges for chemistry education in Africa and possible options meant for further exploration.

Thohoyandou, South Africa — Liliana Mammino

Groningen, The Netherlands — Jan Apotheker

Acknowledgements

The Editors would like to express their gratitude to:

- The Committee on Chemistry Education of IUPAC (International Union of Pure and Applied Chemistry) and the Federation of African Chemical Societies, for inviting Prof. L. Mammino to organize ACRICE-2, hosting it at the University of Venda.
- The International Union of Pure and Applied Chemistry (IUPAC), for recognizing the conference as an IUPAC conference and for offering sponsorship meant for young African Researchers.
- The authorities of the University of Venda, who approved hosting the conference and supported its organization: the Dean of the School of Mathematical and Natural Sciences (Prof. Natasha Potgieter), the then Director of International Relations (Mr Cornelius Hagenmeier), and the DVC Academic (Prof. Jan Crafford).
- The external sponsors who provided financial support for the conference: the American Chemical Society, the Royal Society of Chemistry, the Russian company PhosAgro.
- All the participants, who have made the conference active and lively, sharing their precious expertise and concrete experiences.

Contents

1 **Globalization of Chemistry Education in Africa: Challenges and Opportunities** 1
Mei-Hung Chiu and Yuh-Ru Yu

2 **Achieving the Aims of Practical Work with Microchemistry** 23
John Bradley

3 **Chemistry for the Masses: The Value of Small Scale Chemistry to Address Misconceptions and Re-establish Practical Work in Diverse Communities** 31
Marié H. du Toit

4 **The Systemic Approach to Teaching and Learning Organic Chemistry (SATLOC): Systemic Strategy for Building Organic Chemistry Units** 57
Ameen F. M. Fahmy

5 **Are Our Students Learning and Understanding Chemistry as Intended? Investigating the Level of Prior Knowledge of UNIVEN Students for the Second Year Inorganic Chemistry Module** 69
Malebogo A. Legodi

6 **Are the Newly Formed Kenyan Universities Ready to Teach Externally Examined Diploma Courses in Analytical Chemistry?** 85
Warren A. Andayi

7 **"Closing the Circle" in Student Assessment and Learning** 97
Francis Burns

8 **Interpretation and Translation of Chemistry Representations by Grade 11 Pupils in the Chipata District, Zambia** 113
Lawrence Nyirenda and Salia M. Lwenje

9 The Project IRRESISTIBLE: Introducing Cutting Edge Science into the Secondary School Classroom ... 129
Jan Apotheker

10 Chemistry Teaching and Chemical Education Research: 30-Year Experience in Integration Pathways ... 143
Liliana Mammino

11 Teaching Modern Physics to Chemistry Students ... 161
Joseph K. Kirui and Lordwell Jhamba

12 Learning About Green Chemistry Independently: Students' Point of View ... 169
Charles M. Kgoetlana, Hlawulani F. Makhubele, Lemukani E. Manganyi, Emmanuel M. Marakalala, Shirley K. Sehlale, Derrick O. Sipoyo, Neani Tshilande, and Thembani S. Vukeya

13 The African Context: Investigating the Challenges and Designing for the Future ... 177
Liliana Mammino

About the Editors and Contributors

About the Editors

Liliana Mammino is Professor Emeritus in the School of Mathematical and Natural Sciences of the University of Venda (UNIVEN, South Africa). She completed her 5-year chemistry degree at the University of Pisa (Italy) and her PhD at Moscow State University (Russia), with theoretical chemistry theses in both cases. Her work experience includes working in the Departments of Chemistry at the National University of Somalia (1974–1975), University of Zambia (1988–1992), National University of Lesotho (1993–1996) and UNIVEN since 1997, teaching general chemistry and physical chemistry courses and, at UNIVEN, also the process technology course. Her research interests comprise theoretical/computational chemistry (with particular focus on the computational study of biologically active molecules) and chemistry education (with particular focus on conceptual understanding, the role of language in science learning, and green chemistry education). She has published extensively in both areas. She is a titular member in Division III of IUPAC and representative of Division III at the IUPAC Interdivisional Committee on Green Chemistry for Sustainable Development. She was a recipient of the IUPAC award for *Distinguished Women in Chemistry and Chemical Engineering* in 2013 and of an *ALASA (African Language Association of Southern Africa)* award for sociolinguistics in 2016.

Jan Apotheker recently retired from the University of Groningen as a lecturer in Chemistry Education. He has been involved in the introduction of context-oriented chemistry in the Netherlands, and coordinated a European Project on the introduction of Responsible Research and Innovation. He published two books on chemistry education. He is the current chair of the Committee on Chemistry Education of IUPAC.

Contributors

Warren Andrew Andayi is a chemistry lecturer and Dean of students at the Murang'a University of Technology (Kenya). He holds a PhD in synthetic and medicinal chemistry from the University of Cape Town (2011), an MSc in chemistry from Maseno University (2005) and a BEdSc (mathematics & chemistry) from Kenyatta University (1997). He is a registered teacher and has taught in high school for over 5 years (1998–2007), in addition to teaching and tutoring/demonstrating at Universities in Kenya and South Africa (2007–present). His postdoctoral research work was in tuberculosis drug discovery, liver toxicity diagnostics development, and malaria drug discovery at the University of Cape Town (2011–2012), Rhodes University, Grahamstown (2012–2013) and University of Pretoria (2013–2014), respectively. He has presented his research work in international conferences and published in reputable peer-reviewed journals such as *ACS Med Chem Lett*, *Malaria Journal*, *Bioorg Med Chem Lett*, *Bioorg. Med. Chem*. His current research interests are in chemical education, STEM education and medicinal chemistry.

John Bradley began his career in South Africa as a lecturer in chemistry, but quite soon modified his career track to chemistry education. The challenges in this field were, and are, huge, as limited funds and teachers combine to make it so. Based throughout his career at Wits University in Johannesburg, he has been able to engage in teaching and research towards these challenges. He took a leading role in the University Academic Support Programme, initiated the creation of a School of Science Education and founded the RADMASTE (Research and Development in Mathematics, Science and Technology Education) Centre. In the course of this work, he participated internationally through the UNESCO Global Programme in Microscience and chaired the then Committee on Teaching of Chemistry of IUPAC (now called Committee on Chemistry Education). Within South Africa, he has served in various capacities, such as President of the SA Chemical Institute and Chairman of the Ministerial Task Team on MST Education.

Francis Burns holds a BS in entomology and an MA in science education from Ohio State University (Columbus, OH, USA), and a PhD in chemistry from the University of Toledo, Ohio. He has taught at various American universities and colleges: University of Findlay (Ohio), Grand Valley State University (Michigan), Ferris State University (Michigan), Diné College (Arizona), and the University of South Carolina at Salkehatchie. In addition to teaching at American institutions, he has lectured analytical chemistry and performed mercury analysis at the University of Koya (Kurdistan, Iraq). He mentored a small group of Navajo Native Americans through a high-altitude atmospheric chemistry research project, which was sponsored by NASA. While being trained in the analytical sciences, he is also interested in the variables affecting student success, particularly linguistic barriers and psychological factors affecting student engagement. His current institution (USC

Salkehatchie) has a deep commitment to students seeking to gain an advanced education, but lacking requisite skills from their prior schooling.

Mei-Hung Chiu is a Professor of Science Education at the Graduate Institute of Science Education of the National Taiwan Normal University (NTNU). She taught chemistry in Taiwan before completing her doctorate in Harvard. Her research interests include (1) better understanding and promoting students' conceptual constructions and changes in sciences; (2) developing students' understanding of scientific models and modelling processes, and (3) exploring whether facial micro-expression state (FMES) changes can be used to identify moments of conceptual conflict scenarios. Chiu has published more than 100 peer-reviewed journal articles (in English and Chinese) and four edited books (in English). She was the President of the National Association for Research in Science Teaching (NARST, 2016-2017) based in the USA and also served as associate editor of the *Journal of Research in Science Teaching* (JRST) for five years. She has chaired the IUPAC Committee on Chemistry Education for four years. She was a recipient of the Distinguished Contribution to Chemical Education Award from the Federation of Asian Chemical Societies (FACS) in 2009 and a recipient of the Distinguished Contribution to Science Education Award from Eastern-Asian Science Education Association (EASE) in 2016.

Marié Henriette du Toit holds an MSc and HED with a specialization in physical science and mathematics from the North-West University, Potchefstroom (South Africa). She taught physical sciences and mathematics for a number of years at the Bellville Technical High School. During a three-year period in the United Kingdom, she worked as a physics research assistant at the Cavendish Laboratory in Cambridge. After teaching chemistry for a number of years and presenting specialized classes and winter schools, she became one of the founder members of the SEDIBA in-service physical science teacher training project at the Potchefstroom campus of the North-West University. For 19 years, she has trained both in-service and pre-service educators in chemistry and chemical education. In the last three years, she has been a chemistry lecturer for the general chemistry course for B. Ing engineering students. Her main research interest is practical work in chemistry. She has been the co-leader of the MYLAB small-scale chemistry and natural sciences project for 22 years. She wrote the CAPS document for chemistry in South Africa (the Curriculum Assessment Policy Statement for the new National Curriculum Statement for each school grade in South Africa).

Ameen Farouk Mohamed Fahmy was appointed Professor of Organic Chemistry at Ain Shams University (ASU), Cairo, Egypt in 1979 and is Professor Emeritus since 2003 to date. He has supervised about 60 MSc and PhD students in organic chemistry. In 1976, he was appointed as a chemistry expert in the Science Education Centre (SEC) of ASU and in 2000–2003 he was the director of SEC. In 1998, Ameen Fahmy and Professor J. J. Lagowski, started a collaboration between ASU and the University of Texas in Austin (USA) on an innovative approach for teaching and learning which they called the Systemic Approach to Teaching and Learning (SATL, website: www.satlcentral.com). Fahmy has about 150 publications in his two areas of expertise, Chemistry and Chemistry Education. He serves as a member of different international, American, European and African Societies, Committees and Associations. He was invited as PL/keynote speaker in different international, Arab and African conferences. He has organized about 90 International, National, and Arab Conferences, Workshops, Seminars, and training programs on SATL applications, Systemic Assessment and TQ-programs in Education, and was chairperson of the 15th ICCE [Cairo, August1998]. He also acts as a visiting professor in different European, American, African and Arab Universities.

Lordwell Jhamba holds a PhD in Physics from the University of the Witwatersrand, as well as an MSc, MScEd, BSc (Hons) and BEd (Phy) from the University of Zimbabwe. He is a senior lecturer in the Department of Physics of the University of Venda. He was a lecturer in the Departments of Physics of Midlands State University (2004–2007) and Hillside Teachers' Associate College of the University of Zimbabwe (1990–2001), and in the Departments of Physics and Physics Education in the Bindura University of Science Education (2002–2003), all in Zimbabwe. His research interests comprise condensed matter and materials physics (experimental), renewable energy (photovoltaic, wind, solar, thermal and biomass systems), and science education (physics, mathematics for physics, physical science).

Charles M. Kgoetlana, Hlawulani F. Makhubele, Lemukani E. Manganyi, Emmanuel M. Marakalala, Shirley K. Sehlale, Derrick O. Sipoyo, Neani Tshilande and Thembani S. Vukeya were third year students at the University of Venda in 2015, taking the process technology course in the second semester. They all passed the course in the November exam. Most of them are currently continuing with postgraduate studies (Marakalala, Sipoyo, Tshilande and Vukeya at the University of Venda, Selahle at the University of Johannesburg); others are working as professionals.

Joseph Kiprono Kirui obtained a BSc (Maths and Physics) in 1984 at the University of Nairobi, Kenya and was appointed a lecturer at Kenya Polytechnic, Nairobi. He obtained his MSc (Physics) from the University of British Columbia, Vancouver (Canada) in 1990 and was appointed assistant lecturer at Egerton University, Njoro (Kenya) the same year. He was appointed as lecturer at the University of Venda (UNIVEN), Thohoyandou (South Africa) in 2002, while he was studying for his PhD (Physics) at the University of the Witwatersrand, Johannesburg, where he graduated in 2008. He is a senior lecturer in the Department of Physics at UNIVEN, where he has been teaching modern physics, waves and optics, electromagnetism and solid-state physics for undergraduates since 2002, and solid-state physics at honours level since 2013 to date. He was head of the physics department and member of the Senate of UNIVEN in the years 2009–2017. He has been an associate member of the National Institute of Theoretical Physics (NITheP) of South Africa and served in its steering committee since 2015. Kirui has promoted/supervised five postgraduate (MSc and PhD) students to completion and has published eight articles in peer-reviewed journals.

Malebogo Andries Legodi is Senior Lecturer of Materials and Inorganic Chemistry at the University of Venda, Thohoyandou, South Africa. He holds a BSc in chemistry from the University of Cape Town and an MSc and PhD in inorganic chemistry from the University of Pretoria (UP), South Africa. He also holds (with a distinction) a postgraduate diploma in Higher Education Studies (Teaching and Learning) from Rhodes University, South Africa. He has previously worked as a lecturer at UP and as postdoctoral fellow in the Material Science and Manufacturing Business Unit of the Centre for Scientific and Industrial Research (CSIR) in Pretoria. His research interests include the study of complexes with metal atoms or ions and the synthesis and characterization of bulk and nano materials of inorganic compounds for use as cathode materials in lithium ion batteries, as well as for medicinal purposes. He also has keen interest in research related to chemical education in higher education.

Salia Magdalene Lwenje, until her death on 29th November, 2018, was a senior lecturer in the Chemistry Department, Copperbelt University, Kitwe, Zambia. At the time of her death, she was also Director – Academics for the newly established Mukuba University within Kitwe, which, in its transitional period, was run under Copperbelt University caretaker administration. She held a BSc (chemistry/mathematics) from the University of Zambia and an MA and PhD in chemistry from Brandeis University, USA. On completion of her BSc, she was awarded a Staff Development Fellowship by the University of Zambia, which paid for her postgraduate studies at Brandeis University. Her work experience included teaching chemistry at the University of Zambia (1985–1989) and at the University of Swaziland (1990–2013). Her research interests comprised the investigation of heavy metals in the environment, the conversion of organic waste to biogas, investigating indigenous fruits as sources of oil for potential use as a biofuels, aluminum toxicity in acidic soils, and chemistry education. She authored/coauthored 9 papers in refereed journals and a number of scientific reports, and contributed chapters to two books. She died in a road traffic accident as she travelled to Lusaka, the nation's capital, on official duties.

Lawrence Nyirenda is Senior Education Standards Officer for Natural Sciences in the Zambian Ministry of General Education. He holds a degree in chemistry from the University of Zambia and an MSc in chemistry education from the Copperbelt University. In 2015, he was awarded the Copperbelt University award for the best graduating student in MSC in chemistry education. His work experience includes being a teacher of science at secondary school, deputy head-teacher, and senior lecturer in a college of education. His research interests comprise the study of the interpretation and translation of chemistry representations by pupils at secondary school in Zambia. He is currently conducting research on the use of particulate diagrams in teaching secondary school chemistry; this research is intended to be part of PhD studies to be undertaken. Throughout his teaching career, he has taken part in organizing and facilitating activities in the Zambia Association of Science Educators (ZASE), meant to improve the teaching and learning of all science subjects, and in the Junior Engineers, Technicians and Scientists (JETS) of Zambia, which promotes innovations in the use of concepts in various science subjects by learners, at all levels of the Zambian education system.

Yuh-Ru Yu is an adjunct assistant professor in the College of Business and Management, Tamkang University, Taiwan. She holds a PhD from the Department of Management Sciences in Tamkang University, Taiwan, in 2010. Her main research interests are in statistical computing and simulation, data mining, ranking and selection procedure, decision theory, and multiple decision procedures. She was a data analyst in Prof. Mei-Hung Chiu's Lab of Conceptual Change and Modeling in Science Education, Graduate Institute of Science Education, National Taiwan Normal University, Taiwan, during 2011–2019, and has published articles in international journals.

Chapter 1
Globalization of Chemistry Education in Africa: Challenges and Opportunities

Mei-Hung Chiu and Yuh-Ru Yu

1 Introduction

Cultivating scientific literacy across a country's citizenry is necessary today in order to enjoy nationwide economic and societal development (Chiu and Duit 2011; Hodson 2003). The push for scientific literacy is evidenced on a global scale, with countries being concerned they will become isolated or marginalized if they fail to enact science education reform in terms of curriculum standards, innovative teaching strategies, and professional development. Underlying each aspect of this reform movement is the creation of student-centred learning environments.

As for curriculum standards, there is a shifting away from "learning by doing", "science for all", and "less is more" to the current Next Generation Science Standards (NGSS), which encompass three perspectives, namely disciplinary core ideas (DCIs), crosscutting concepts, and science and engineering practices. DCIs are designed to highlight the key ideas in science, both within a single discipline and across multiple disciplines in science and through various grade levels. More importantly, the DCIs are related to students' daily life experiences and personal and societal concerns in order to arouse students' interest in learning science. Crosscutting concepts are intended to help students identify and explore connections across science domains and develop coherent and scientifically-based views of the world around them. Through the link between DCIs and crosscutting concepts, students develop their ability to use causal reasoning, evaluate and use evidence, engage in argument with evidence, and develop models in science learning. The science and engineering practices provide students with the opportunity to investigate the natural world, design and build systems like scientists and engineers do, and deepen their understanding by applying their knowledge of core ideas and

M.-H. Chiu (✉) · Y.-R. Yu
Graduate Institute of Science Education, National Taiwan Normal University, Taipei City, Taiwan
e-mail: mhchiu@gapps.ntnu.edu.tw

L. Mammino, J. Apotheker (eds.), *Research in Chemistry Education*,
https://doi.org/10.1007/978-3-030-59882-2_1

crosscutting concepts. The term "practices" was chosen over "skills" to emphasize the cognitive, physical, and social practices that are inherent to scientific inquiry (NRC 2013).

The NGSS are utilized in many different countries and considered an influential reference for science curriculum and science education reform. The global adoption of the NGSS appears to be due to the internationally shared recognition that the scientific enterprise that scientists experience should be shared with students in order to help students develop sophisticated scientific understanding of key ideas of science, to appreciate what science is about and what science is for, and to be able to adopt science information into their daily lives. In addition, engaging students in scientific practices, such as modelling, helps them construct, evaluate, and communicate scientific knowledge in a systematic manner.

2 Results from International Monitoring Systems

The two largest international monitoring systems for students' science performance and scientific literacy, TIMSS (Trends in International Mathematics and Science Study) and PISA (Programme for the International Student Assessment), play an influential role in international science education reform. Based on these assessments, a number of countries (e.g., Denmark and Germany) investigated their science education systems and programs and then shifted the direction of their reform efforts in science education (Dolin and Krogh 2010; Neumann et al. 2010). A country's international standing (and status) in terms of its students' science and mathematics performance is used to assess national educational policies and is a barometer of whether students are learning necessary knowledge and skills to be successful locally and globally (Yore et al. 2010).

While over 60 countries across the world participate in the PISA, only three countries in Africa joined the study over the past 15 years. Tunisia was the first to join in 2003, Mauritius in 2009 (Organisation for Economic Co-operation and Development (OECD) 2008), and Algeria began participating in 2015. As for TIMSS, six African countries (Algeria, Botswana, Egypt, Morocco, South Africa, and Tunisia) participated in different years. South Africa was the first African nation to participate in TIMSS in 1995, followed by Tunisia and Morocco in 1999.

Tables 1.1, 1.2, and 1.3 show the science results of TIMSS 4th grader, TIMSS 8th grader, and PISA (15-year olds) for years between 2003 and 2012 (Martin et al. 2004, 2008, 2012). The tables' columns are students' averaged score (Mean), boys' averaged score (Boys), girls' averaged score (Girls) and ranking (Rank) of the above years' international assessments. All the information in the tables, except Rank, was drawn from the assessment reports released by the International Association for the Evaluation of Education Achievement (IEA) and the Organization for Economic Co-operation and Development (OECD). The authors prepared the ranking shown in the tables based on the participant countries' average score in the assessment reports. The three tables show the performances of all African participant countries

Table 1.1 TIMSS-Science (4th graders) in 2003, 2007, and 2011

	2003				2007				2011			
	(*N* = 25)				(*N* = 36)				(*N* = 52)			
Country	Mean	Boys	Girls	Rank	Mean	Boys	Girls	Rank	Mean	Boys	Girls	Rank
Singapore	565	565	565	1	587	587	587	1	583	585	581	2
Chinese Taipei	551	**555***	548	2	557	558	556	2	552	**555***	548	6
Korea	–	–	–	–	–	–	–	–	587	**590***	583	1
Japan	543	545	542	3	548	547	548	4	559	561	556	4
Hong Kong	542	541	544	4	554	556	553	3	535	**538***	532	9
England	540	538	542	5	542	540	543	7	529	528	529	15
Finland	–	–	–	–	–	–	–	–	570	570	570	3
United States	536	**538***	533	6	539	541	536	8	544	**549***	539	7
Germany	–	–	–	–	528	**535***	520	12	528	**534***	522	17
International average	**489**	**488**	**489**		**500****	**474**	**477***		**500****	**485**	**487**	
Algeria	–	–	–	–	354	349	**359***	31	–	–	–	–
Botswana	–	–	–	–	–	–	–	–	367	360	**374***	***
Tunisia	314	312	316	24	318	304	**335***	33	346	334	**359***	48
Morocco	304	303	306	25	297	292	302	34	264	259	**268***	49

Note: 1. *Gender difference statistically significant. 2. **TIMSS scale centerpoint. 3. ***Sixth grade participants. 4. Country order mainly based on TIMSS 2003 country ranking order. 5. N: stands for numbers of the participating countries

Table 1.2 TIMSS-Science (8th graders) in 2003, 2007, and 2011

	2003				2007				2011			
	(*N* = 46)				(*N* = 49)				(*N* = 45)			
Country	Mean	Boys	Girls	Rank	Mean	Boys	Girls	Rank	Mean	Boys	Girls	Rank
Singapore	578	579	576	1	567	563	571	1	590	591	589	1
Chinese Taipei	571	572	571	2	561	563	559	2	564	564	564	2
Korea	558	**564***	552	3	553	**557***	549	4	560	563	558	3
Hong Kong	556	**561***	552	4	530	528	533	9	535	534	536	8
Japan	552	**557***	548	6	554	556	552	3	558	**562***	554	4
Finland	–	–	–	–	–	–	–	–	552	550	555	5
United States	527	**536***	519	9	520	**526***	514	11	525	**530***	519	10
England	544	**550***	538	****	542	546	537	5	533	532	534	9
International average	**474**	**477***	**471**		**500****	**463**	**469***		**500****	**474**	**480**	
Algeria	–	–	–	–	408	408	408	42	–	–	–	–
Egypt	421	421	422	35	408	400	**417***	41	–	–	–	–
Tunisia	404	**416***	392	38	445	**455***	436	34	439	**447***	431	29
Morocco	396	**403***	392	40	402	401	403	****	376	374	378	41
Botswana	365	366	364	43	355	343	**365***	46	404	399	**410***	***
South Africa	244	244	242	45	–	–	–	–	332	328	335	***

Note: 1. *Gender difference statistically significant. 2. **TIMSS scale centerpoint. 3. ***Ninth grade participants. 4. ****Did not satisfy guidelines for sample participation rates. 5. Country order mainly based on TIMSS 2003 country ranking order. 6. N: stands for the number of participating countries

Table 1.3 Students' performance on PISA in 2003, 2006, 2009, and 2012

	2003				2006				2009				2012			
	$N = 41$ OECD countries: 30 Partner countries: 11				$N = 57$ OECD countries: 30 Partner countries: 27				$N = 65$ OECD countries: 34 Partner countries: 31				$N = 65$ OECD countries: 34 Partner countries: 31			
Country	Mean	Boys	Girls	Rank	Mean	Boys	Girls	Rank	Mean	Boys	Girls	Rank	Mean	Boys	Girls	Rank
Finland	548	545	**551***	1	563	562	565	1	554	546	**562***	2	545	537	**554***	5
Shanghai-China	–	–	–	–	–	–	–	–	575	574	575	1	580	583	578	1
Singapore	–	–	–	–	–	–	–	–	542	541	542	4	551	551	552	3
Japan	548	550	546	2	531	533	530	6	539	534	545	5	547	**552***	541	4
Hong Kong-China	539	538	541	3	542	546	539	2	549	550	548	3	555	558	551	2
Chinese Taipei	–	–	–	–	532	536	529	4	520	520	521	12	523	524	523	13
Korea	538	**546***	527	4	522	521	523	11	538	537	539	6	538	539	536	7
Germany	502	506	500	18	516	519	512	13	520	523	518	13	524	524	524	12
United Kingdom	518	**	**	**	515	**520***	510	14	514	**519***	509	16	514	**521***	508	21
OECD average	**500**	**503**	**497**		**500**	**501**	**499**		**501**	**501**	**501**		**501**	**502**	**500**	
United States	491	494	489	22	489	489	489	29	502	**509***	495	23	497	497	498	28
Tunisia	385	380	**390***	40	386	383	388	54	401	401	400	57	398	399	398	61

Note: 1. *Gender difference statistically significant. 2.**Response rate too low to ensure comparability. 3. Mauritius participates in PISA in 2010 not in 2009. Please see the info at https://www.oecd.org/pisa/aboutpisa/pisa2009participants.htm. 4. Country order mainly based on PISA 2003 country ranking order. 5. N: stands for the numbers of OECD countries and partner countries from 2003 to 2012 and also include scores from selected Asian, European and American participant countries.

For the sake of comparison, the performances of fourth graders from participating African countries, along with selected other countries, are shown in Table 1.1. As one can see, the top four countries in 2003 and 2007 were all from Asia. Table 1.1 also shows that the average scores on TIMSS for fourth graders in 2003, 2007, and 2011 were 489, 500, and 500, respectively. Tunisia scored 314 (ranked 24th), 318 (ranked 33rd), and 346 (ranked 48th), respectively, and Morocco scored 304 (ranked 25th), 297 (ranked 34th), and 264 (ranked 49th) respectively, revealing large gaps between the average TIMSS scores and the individual scores of these two countries (Table 1.1). Algeria and Botswana participated in TIMSS in 2007 and 2011, respectively. Both countries outperformed other African countries, but were still far away from the average scores of all the participating countries. Additionally, with regard to gender difference in science performance in participating African countries, girls' average scores were all higher than those of boys, and the gender differences in scores were statistically significant in Algeria (TIMSS 2007), Botswana (TIMSS 2011), Tunisia (TIMSS 2007 and 2011) and Morocco (TIMSS 2011). As for non-African countries, there was a performance difference between girls and boys, but this difference was not statistically significant in most countries. Interestingly, in cases where non-African countries' gender difference in scores were statistically significant, it was the boys who had the higher average scores (e.g. Chinese Taipei (Taiwan), United States, and Germany). Overall, when the gender difference in performance was statistically significant, girls outperform boys among African countries and the reverse is true among non-African countries. This is especially true in TIMSS 2011.

Similarly, the scores from TIMSS for eighth graders from six African countries revealed low performance in the science field (see Table 1.2). In 2003, Egypt received the highest scores among the African countries. However, Tunisia outperformed Egypt in 2007 and continued improving its rank in 2011. South Africa was the last among them. Perhaps due to financial issues and low student performance in TIMSS, South Africa did not participate in 2007, and Algeria and Egypt did not participate in 2011. Even though South Africa participated again in 2011 and the country's scores increased, the country's rank was considered unchanged since 2003 (see Table 1.2). Regarding gender differences among eighth graders in participant countries, boys performed better overall in TIMSS 2003, since they performed better in most countries. The gender difference was statistically significant in Korea, Hong Kong, Japan, United States, England, Tunisia and Morocco in 2003. In TIMSS 2007 and 2011, girls performed better among African countries (except Tunisia), while boys performed better in Tunisia and in some of the non-African countries listed in the table. Overall, when gender difference was statistically significant, girls performed better in science than boys among African countries (mainly in Egypt and Botswana, excluding Tunisia) and boys performed better in Tunisia and in several non-African countries (mainly Korea, Japan, and the United States).

Table 1.3 shows the scientific literacy of 15-year-old students from selected countries participating in PISA in 2003, 2006, 2009, and 2012 (Organisation for

Economic Co-operation and Development (OECD) 2004, 2007, 2010, 2014). The top countries were Finland, Japan, Hong Kong, and Korea in 2003, while Shanghai was ranked 1 in 2009 and 2012, when it participated in PISA. However, Tunisia was about 100 points below the average PISA scores. These results are consistent with the information presented in Tables 1.1 and 1.2. Regarding the gender difference in terms of science performance, this difference is statistically significant for Finland, where girls outperformed boys (PISA 2003, 2009, 2012), and for UK, where the reverse was true (PISA 2006, 2009, 2012). In the majority of countries, irrespective of year, there is a slight performance difference between the genders, but the difference is not statistically significant, and there is no observable pattern of either gender outperforming the other. As for the only African country included in Table 1.3, Tunisia, girls outperformed boys in all the years considered (2003, 2006, 2009, 2012), but the difference was statistically significant only in 2003.

However, additional PISA data shed light on how to support science education in Africa. Tables 1.4 and 1.5 show that several countries (e.g., Finland, Hong Kong, Japan, Korea, and Taiwan) with high performance in scientific literacy, also demonstrated low interest in chemistry/science and low levels of self-efficacy in school. Interestingly, students in Tunisia reported relatively high levels of interest in the following categories: chemistry, the ways scientists design experiments, what is required for scientific explanations, and wanting to learn science. These scores were not only above the OECD average scores, but Tunisia's scores were also far above those of high-performing countries in other continents. In particular, Tunisia's score for "enjoyment of science" (see Table 1.5) and "motivation to learn science" (see Table 1.6) were astonishingly high compared to other countries who received high

Table 1.4 Index of general interest in science

Country	C. Topics in chemistry	F. Ways scientists design experiments	H. What is required for scientific explanations
Finland	45	24	26
Hong Kong	55	53	44
Chinese Taipei	46	51	42
Japan	48	34	25
Korea	42	24	28
Germany	59	54	42
United Kingdom	55	41	35
OECD average	**50**	**46**	**36**
United States	56	45	34
Tunisia	67	72	64

Percentage of students reporting high or medium interest in the issues stated in the columns' headings

Note: 1. Extracted from Organisation for Economic Co-operation and Development (OECD) (2007) Figure 3.8. 2. Singapore and Shanghai-China did not participate in PISA in 2006. 3. Country order mainly based on PISA 2006 country ranking order in Table 1.3

Table 1.5 Index of enjoyment of science

Country	A. I enjoy acquiring new knowledge in science	B. I generally have fun when I am learning science topics	C. I am interested in learning about science	C. I like reading about science	D. I am happy doing science problems
Finland	74	68	68	60	51
Hong Kong	85	81	77	65	54
Chinese Taipei	79	65	64	62	43
Japan	58	51	50	36	29
Korea	70	56	47	45	27
Germany	52	63	60	42	38
United Kingdom	69	55	67	38	53
OECD average	**67**	**63**	**63**	**50**	**43**
United States	67	62	65	47	41
Tunisia	95	87	91	85	76

Percentage of students agreeing or strongly agreeing with the statements in the columns' headings
Note: 1. Extracted from Organisation for Economic Co-operation and Development (OECD) (2007) Figure 3.10. 2. Singapore and Shanghai-China did not participate in PISA in 2006. 3. Country order mainly based on PISA 2006 country ranking order in Table 1.3

scores on science performance in academia. These findings were surprising and exciting to Western and Asian researchers as well as African researchers (Chiu 2015b). Although Tunisia's data cannot be generalized to all African countries, Tunisian students' positive attitude toward science needs to be acknowledged and built on. It is important to build on the momentum from TIMSS and PISA to further improve the quality of science/chemistry education, science/chemistry teachers, and learning resources available to African students (see Tables 1.4, 1.5, and 1.6).

Aside from the academic performance shown in Table 1.3, the PISA 2006 results (see Table 1.7) showed that Tunisian students had low awareness of environmental issues that have relevance to daily life. The great gap between Tunisia's scores regarding environmental issues and the OECD average scores was surprising. Interestingly, most of the countries did poorly on the item, "Use of genetically modified organisms (GMOs)" except Taiwan (referred to as Chinese Taipei).

From the comparisons shown in Tables 1.1, 1.2, 1.3, 1.4, 1.5, 1.6 and 1.7, it is clear that several challenges exist for African countries. First, lower scores in international comparisons may discourage African countries from joining such international studies. Also, the financial demands of joining an international monitoring system, particularly during times of economic turmoil, may discourage African countries from participating. However, participation in international monitoring systems allows governmental agencies to more thoroughly and regularly assess the development and effectiveness of their science education policies and curriculum

Table 1.6 Index of instrumental motivation to learn science

Country	A. I study school science because I know it is useful for me	B. Making an effort in my school science subject(s) is worth it because this will help me in the work I want to do later on	C. Studying my school science subject(s) is worthwhile for me because what I learn will improve my career prospects	D. I will learn many things in my school science subject(s) that will help me get a job	E. What I learn in my school science subject(s) is important for me because I need this for what I want to study later on
Finland	63	53	51	48	43
Hong Kong	72	73	72	64	63
Chinese Taipei	83	76	76	73	65
Japan	42	47	41	39	42
Korea	55	57	52	46	45
Germany	66	58	55	50	48
United Kingdom	75	71	71	65	54
OECD average	**67**	**63**	**61**	**56**	**56**
United States	77	78	70	70	68
Tunisia	89	89	85	84	86

Percentage of students agreeing or strongly agreeing with the statements in the columns' headings
Note: 1. Extracted from Organisation for Economic Co-operation and Development (OECD) (2007) Figure 3.12. 2. Singapore and Shanghai-China did not participate in PISA in 2006. 3. Country order mainly based on PISA 2006 country ranking order in Table 1.3

standards, in order to more effectively shape the direction of their science education reforms. Second, promising findings exist in students' high scores regarding their attitude toward science, which should be explored as an avenue for chemistry education reforms. If teachers can take advantage of this phenomenon, the school climate would likely be changed for the better. Third, environmental issues should be strengthened in school chemistry teaching, as indicated by the very low awareness of these issues compared to other countries participating in the international monitoring systems. When teachers and students pay more attention to environmental issues, they tend to be more motivated to learn chemistry as a means to improve their standard of living and the state of the planet for future generations. Finally, the results point to the need to promote female participation in the sciences. According to the UN's (2015) Global Goals for Sustainable Development report, women comprise two-thirds of the world's illiterate population. Across Africa, like many other parts of the world, females receive lower wages and attain lower-level employment compared to their male counterparts. How can more women be kept in school and provided with quality education, so they can make a better life for themselves, their

Table 1.7 Index of students' awareness of environmental issues

Country	A. The consequences of clearing forests for other land use	B. Acid rain	C. The increase of greenhouse gases in the atmosphere	D. Nuclear waste	E. Use of genetically modified organisms (GMOs)
Finland	75	60	65	63	22
Hong Kong	91	88	80	48	31
Chinese Taipei	90	84	80	56	54
Japan	68	75	54	33	33
Korea	42	75	53	42	27
Germany	80	65	60	61	38
United Kingdom	74	71	71	59	37
OECD average	**73**	**60**	**58**	**53**	**35**
United States	73	54	53	51	39
Tunisia	64	27	19	32	20

Percentage of students who were familiar with the environmental issues stated in the columns' headings
Note: 1. Extracted from Organisation for Economic Co-operation and Development (OECD) (2007) Figure 3.17. 2. Singapore and Shanghai-China did not participate in PISA in 2006. 3. Country order mainly based on PISA 2006 country ranking order in Table 1.3

families, and their countries? This is a particularly salient issue across most of Africa where less than 50% of the female population completes primary school.

As a closing remark to this section, research can help governments re-evaluate their educational environment and consider strategies and actions to promote more grounded policies and practices. Although the data presented here do not represent the educational status of all African countries, they definitively show that student outcomes in the region lag behind those of most other nations. The impact of international assessment results and research outcomes can inform future educational reforms across the continent.

The next section proposes an initiative entailing the use of innovative technology in chemistry education in African school systems, utilising the rapid growth of technology in the world, and particularly in Africa.

3 The Use of Mobile Technology and Augmented Reality for Chemistry Learning

3.1 Background and Approaches

The UN's (2015) Human Development Report stated that, in 2013, people aged 24 and younger accounted for 42.4% of the world's population but 45% of Internet users. Nearly 81% of households in developed countries had Internet access, compared with 34% in developing countries in 2013. While Internet access is not comparable between developed and developing countries, according to Poushter (2016), smartphone ownership in emerging and developing nations is rising at an extraordinary rate, climbing from a median of 21% in 2013 to 37% in 2015. In addition, there were rapid gains in reported Internet access rates in a large number of emerging and developing nations surveyed since 2013. For example, in 2014, only 38% of Nigerian Internet users said they accessed the Internet several times a day, while that number jumped to 58% in 2015. In other words, there was a dramatic and unexpected increase in Internet users, which may imply an increasing number of persons carrying smartphones in Africa.

According to Adepetun (2015), 67% of the African population (1.13 billion in total), now have mobile phones. Tahabalala (2015), see Fig. 1.1) also confirmed that the number of smartphones in Africa increased sharply and is expected to continue increasing for at least the next few years. Smartphones are and will be an affordable means for people in Africa to connect with others locally as well as globally.

Therefore, there is no doubt that infusing this technology into science classrooms would transform schools and provide teachers with a new kind of teaching tool. As mobile technology becomes ubiquitous across Africa, the use of smartphones for science learning has becomes more possible and appealing.

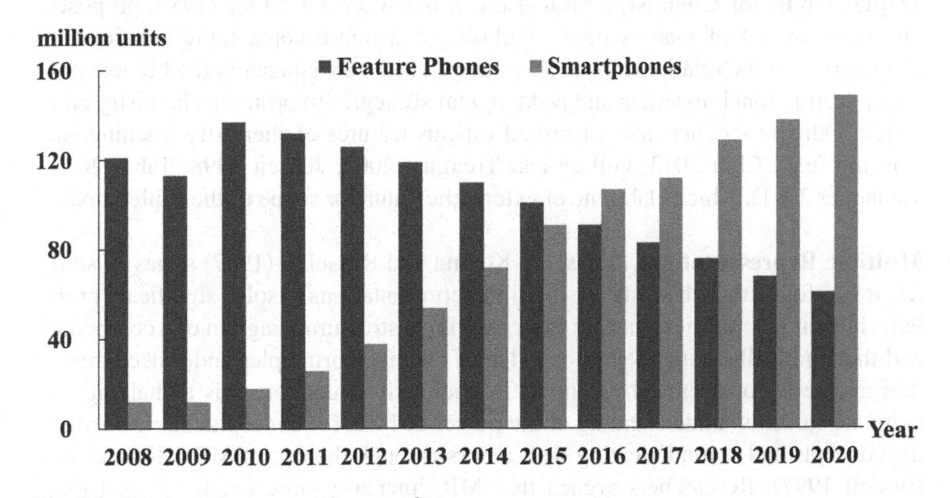

Fig. 1.1 Rise in Africa's smartphone market

In recent years, technology-enhanced learning (TEL), such as augmented reality (AR), and its efficaciousness in promoting students' science learning have been the subject of much research. Central to the TEL model is the placement of students at the center of the learning process, which allows students to be actively engaged in the use of technology (Bacca et al. 2014). Although there are various definitions for AR, it is widely accepted that AR has the ability to bridge virtual and real environments and open up new opportunities for the building of more engaging and constructive learning contexts (Duh and Klopfer 2013). Research in the area of AR shows that the physical component of AR experiences benefits students' learning (Bujak et al. 2013; Ibáñez et al. 2014). Other features of AR linked to successful science learning include: (a) learning content from a 3D perspective; (b) learning that is ubiquitous, collaborative, and situated; (c) sensing presence, immediacy, and immersion; (d) visualizing the invisible; and (e) bridging formal and informal learning (Wu et al. 2013). Although AR can be effective for improving students' learning, learning motivation, engagement, and attitude, the adoption of AR for education (teacher training) was minimally explored in the literature (Bacca et al. 2014). Also, according to Bacca and his colleagues' (Bacca et al. 2014) systemic review, very few software programmes considered students' unique needs in AR, and limited sources are available concerning guidelines for designing AR-based educational settings (p. 144).

A series of newly developed science (e.g., chemistry, biology, physics, and earth science) learning apps featuring AR was designed in the authors' laboratory, based on the following learning keystones: Johnstone's famous triplet model for chemistry, multiple representations, and visualization. The use of multiple representations and interactive devices helps teachers present abstract and complex structures of chemical compounds to improve students' understanding of chemistry concepts in the classroom. The three learning keystones are briefly discussed in the next paragraphs, together with examples of their implementation via AR-based apps.

Triplet Model for Chemistry Education Since when Johnstone (1982) proposed his triplet model of macroscopic, symbolic, and microscopic representations of chemistry, and the relationships among them, chemistry educators tend to use it to set up instructional materials and pedagogical strategies to promote chemistry education. Other researchers also identified various features of chemistry teaching and learning (e.g., Chiu 2012; Gilbert and Treagust 2009; Jensen 1998; Taber 2013; Talanquer 2011), which elaborate or extend the nature or scope of the triplet model.

Multiple Representations (MRs) In Kozma and Russell's (1997) series of studies, it was found that chemists use multiple representations to solve significant problems in their laboratories, starting from drawing a structural diagram of a compound and then gradually using their knowledge of chemical principles and procedures to design a series of chemical reactions. In chemistry education, it is a challenge to cultivate students' understanding of the power of MRs for conceptualizing chemistry concepts and communicating with others through the use of MRs (Kozma and Russell 1997). Researchers argued that MRs increase more cognitive load than simple representations while learning complex and abstract scientific concepts. In order to allow instruction to be effective, research on cognitive load suggests that instruction should aim to decrease extraneous and intrinsic cognitive load and increase germane cognitive load (McElhaney et al. 2015).

Visualization McElhaney et al. (2015) used meta-analysis to synthesize two sets of studies published or conducted since the year 2000 to systematically examine learning outcomes and design features of dynamic visualizations in science. They found that the mean overall effect size was marginally significant in favour of dynamic visualizations and further claimed that longer duration of instruction for complex dynamic visualizations allowed learners to better digest information compared to static visuals. The visualization approach allows users to observe target concepts from different aspects (2D or 3D) and transform them between 2D and 3D perspectives.

3.2 Examples of Augmented Reality Apps

This section considers selected examples of ARs for chemistry education developed in the authors' laboratory under the High Scope project sponsored by the Taiwanese Ministry of Science and Technology, and based on the epistemological approach described in the previous section. As previously stated, the rationale for designing these apps was to help students visualize the structure of compounds, and interaction between them, using the innovative and modern technique of AR and smartphones.

The first example concerns the polarity of water (see Fig. 1.2). The results revealed that the students were enthusiastic to use their smartphones to observe how polar compounds, such as water, form the H-bond and how the electronic cloud generates a repulsive charge in response to oxygen ions. It was also found that students' drawings of the electronic cloud for polarity were much more accurate than before the implementation of AR.

The second example involves the three types of structures of carbon nanotubes (CNT, see the CNT_AR at Youtube, Chiu 2015a), which are allotropes of carbon with a cylindrical nanostructure and are members of the fullerene structural family (See Fig. 1.3). The author's research group developed three types of CNT structures (armchair, zigzag, and chiral structures) to help students visualize the complicated structure of CNTs from different angles (Fig. 1.4).

The third example refers to a collaboration with a chemistry teacher from Taipei Municipal Wanfang High School to design structures of organic compounds on a deck of cards for use with smartphone. If the students aim their smartphones' camera at the cards, 3D objects would appear in their smartphone display (see Fig. 1.3). Students can rotate the cards to see the stereo-structures of the compounds listed on each card, and can observe and compare the different structures of chemical compounds. According to the study presented here, it was found that the students were

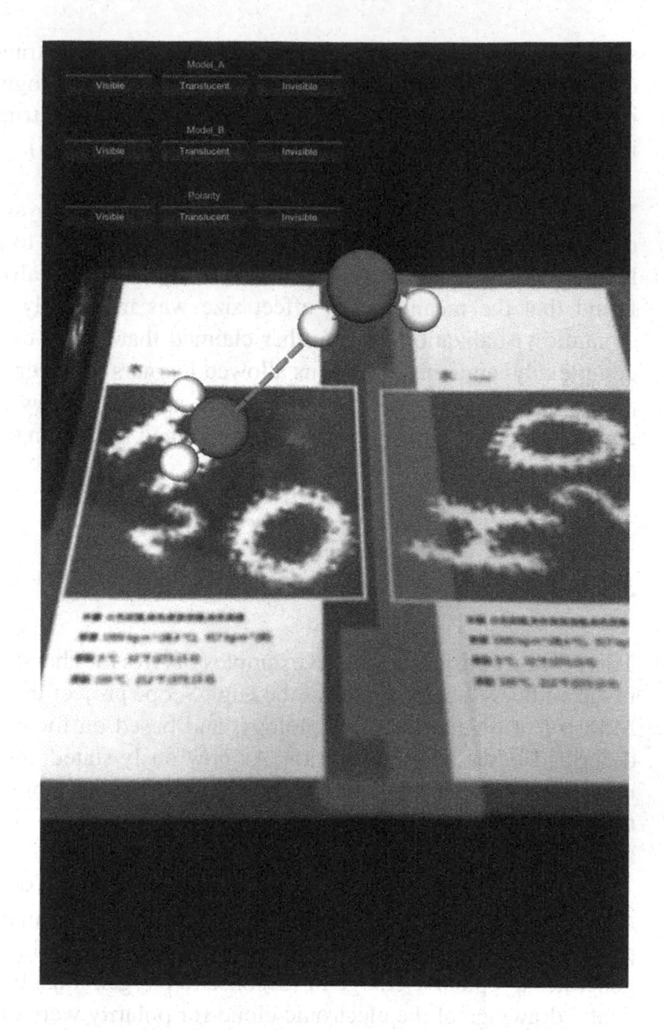

Fig. 1.2 An example of hydrogen bond formation

more motivated to learn chemistry when they utilized the three decks of cards that had AR targets of different types of chemical compounds (such as methane, alkene, and alkyne) in conjunction with their smartphones. Also, the students developed better understanding of the concepts of intermolecular forces among molecules and polarity of a molecule. The app entitled as Molecules AR/VR can be found both in Google Play for Android and Apple Store for IOS (See Fig. 1.5 for examples).

The first author's experiences working with teachers and students over the past 5 years have shown the power of AR, the potential it has to help scaffold students' constructions of chemistry concepts, and the ways it motivates students to learn chemistry. During the second African Conference on Research in Chemical Education, the first author of this chapter also used the apps her research team created to work with African students and teachers. She found that they quickly got used to the apps and were able to switch among different representations of chemical compounds and their symbols to compare the structures (See Fig. 1.6). The

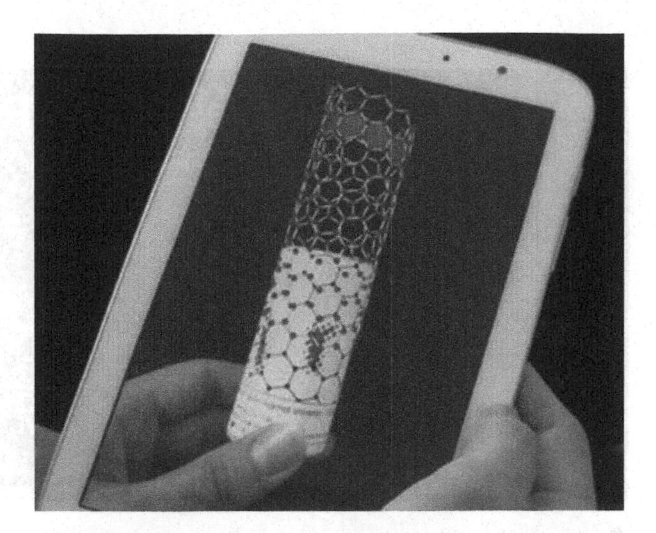

Fig. 1.3 An example of CNT structure

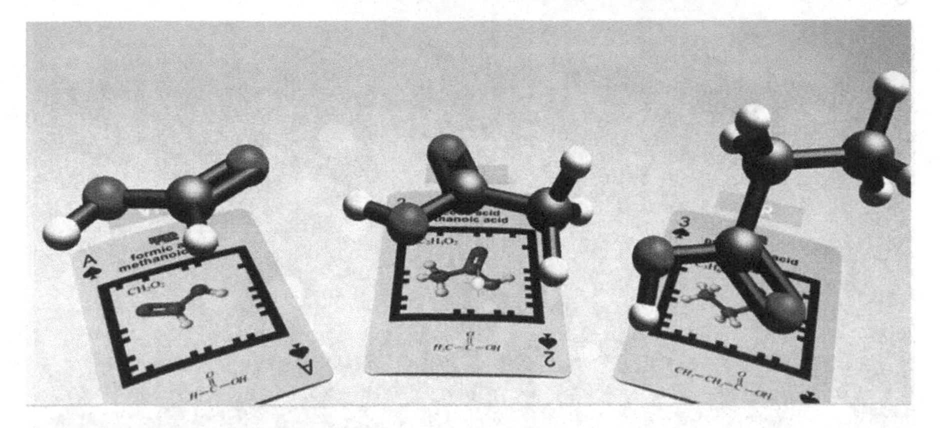

Fig. 1.4 AR for organic compounds

students enjoyed playing by themselves (self-motivated) to find out the relationships among compounds, but also conversed with their classmates about the design and chemical composition of compounds. While the NGSS emphasize the importance of building models for science learning in practice, it is here considered how advanced technology (e.g., smartphones and apps) allows teachers and students to visualize and manipulate models, generate deep understanding of the unseen world of chemistry, and construct their perceptions of the micro-world. The first author also found that teachers played important roles in implementing innovative technology in science teaching. Teachers need to get familiar with the functions of apps, to be able to design appropriate learning activities, and create possible problem solving tasks that are predesigned to allow students to be engaged in science education related tasks.

a

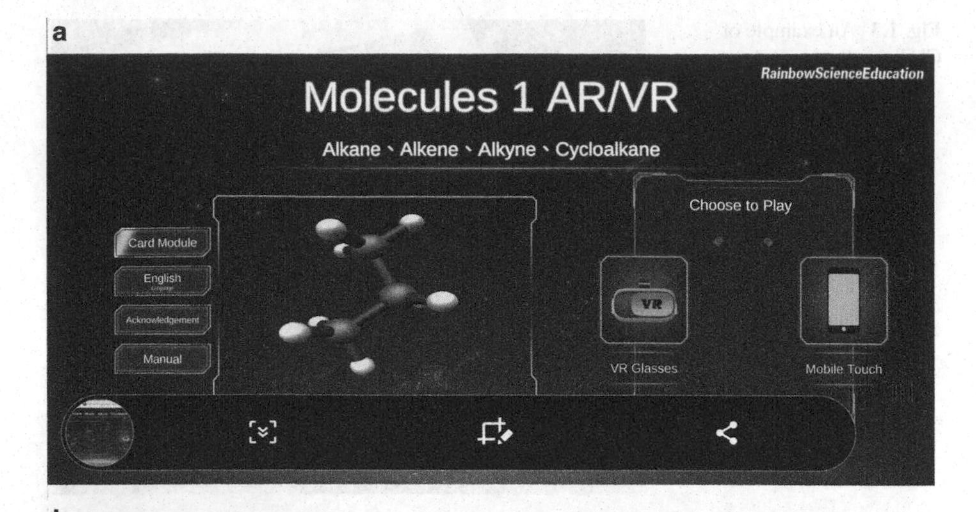

b

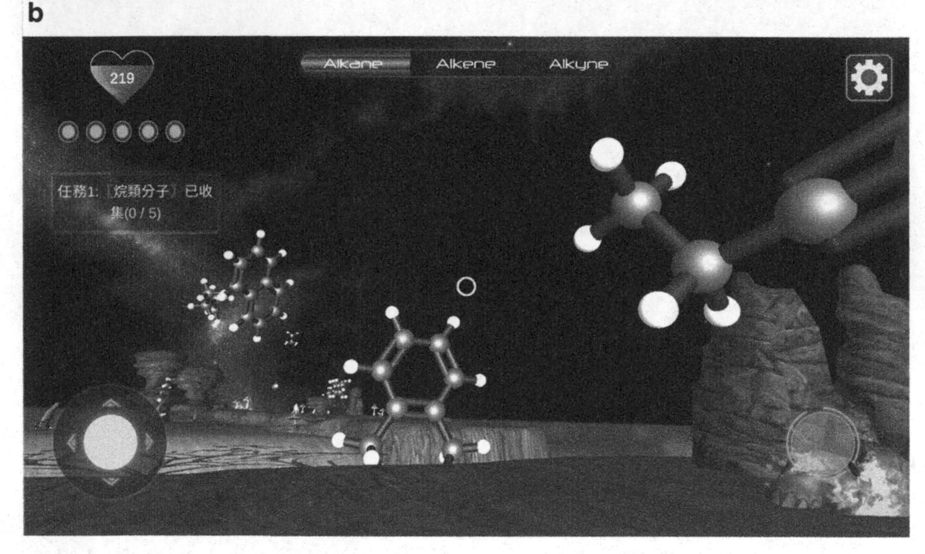

Fig. 1.5 The Molecules AR/VR application

4 Concluding Remarks

Three opportunities are elaborated on in this section. The first opportunity involves using global data from international studies (such as PISA and TIMSS) to inform local policymakers about research-based outcomes and to influence local educational reforms. In particular, participating African countries seemed to show high interest and motivation in learning sciences: why not to take advantage of this positive attitude to enhance their learners' performance in science? The second opportunity capitalizes on the already available instructional e-learning resources and local resources (such as indigenous materials) to stimulate teachers to innovate and

Fig. 1.6 A student from University of Venda using his smartphone to investigate the structure of an organic compound

to implement reforms in their chemistry curriculum development and classroom practices. The authors believe that building a learning community in which secondary and tertiary teachers collaborate and bridge the gap between research and practice in the chemistry classroom can change students' learning motivation and interest in science. This entails using natural phenomena specific to Africa as catalysts to move students from their traditional cultural worldview to a scientific worldview. Third, putting epistemological approaches into curriculum or learning materials, in order to eliminate the gap between macroscopic and microscopic worlds and symbols, should be taken into account. This would allow for the delineation of a formal pathway for developing science knowledge. An example of this in chemistry education is Johnstone's famous triangle that calls for a specific order in the curriculum and gradual development of concepts. In addition, the authors advocate for the inclusion of modeling-based approaches in order to enhance teaching and learning in chemistry by making the invisible visible. In this paper, the authors introduced possibilities for using ARs as scaffolding tools to assist students in understanding MRs of chemical compounds and visualizing those compounds and their structures at different levels of representation. African countries need to perceive the value of science and technology in the current era if they want to have the capacity to use science for the benefit of their people (Chibale 2015). Finally, Ogunniyi (1988) argued that "the pupils' knowledge and image of the world are, to a large extent, determined by the recognition of what their teachers imply to be valid, and this, of course, may have little to do with reality in the sense that it reflects a scientifically accurate picture" (p. 6). With quality teachers in chemistry education, students will be able to develop chemistry literacy and become twenty-first century citizens who would in turn help themselves and their country succeed.

Fig. 1.7 With ACRICE-2 conference chair and her students

Rollnick (1990) described how the many misconceptions associated with scientific concepts found in young children might be rooted in their teachers' misconceptions that they taught formally or informally to their students. In Rollnick et al.'s (2008) case studies of teaching chemistry content about mole and chemical equilibrium in South Africa, they found that although the teachers had some classroom teaching experience and knew their students well, the teachers emphasized procedural (i.e., calculations) rather than conceptual aspects of chemistry concepts, which showed their belief in the importance of teaching for assessment, not for understanding. These studies also pointed out that the teachers lacked understanding of the connections between big ideas and depth of understanding of individual phenomena. This missing link and confidence about subject matter knowledge might explain why their students did not perform well on international monitoring systems of education. Du Toit (2015) used microscale chemical experiments for teaching chemistry for teacher education in order to overcome issues associated with limited resources for doing experiments in schools. She also expressed that in South Africa, many non-chemistry teachers need to teach high school chemistry without having sufficient knowledge and experience in doing experiments. Microscale experiments might be a channel to solve some practical problems. Rollnick et al. (2008) argued that policymakers in South Africa should pay more attention to improving science teachers' subject content knowledge and help them transform that content knowledge into practice.

Final Words from the First Author

My first time in Africa was to attend the 19th annual IUPAC International Conference on Chemistry Education in *Mauritius* in 2008. That was followed by my attending of the celebration of the International Year of Chemistry in Addis Ababa, Ethiopia, in 2011. Both trips provided me with unforgettable experiences and reflections about what chemistry education can do for the people in Africa. In 2015, I was invited to be a plenary speaker for the ACRICE biennial African Conference on Research in Chemical Education, organised by Liliana Mammino, conference chair, at the University of Venda in Thohoyandou, South Africa. The trip to the far north of South Africa allowed me more opportunities to interact with South African students and in-service school teachers at the University of Venda (See Fig. 1.7). Their enthusiasm for sharing and learning impressed and enlightened me in multiple ways. The universal language of science/chemistry and our natural curiosity of the world bring us all closer to each other and highlight our mutual interest in chemistry education and beyond.

Acknowledgements The Ministry of Science and Technology (contract No. 104-2514-S-003-004) and the Ministry of Education, Dr. Chiuthank, for the grants supporting the studies reported in this article. Dr. Liliana Mammino for her invitation to attend the ACRICE conference and deliver a plenary speech, and the many great friends she made across South Africa, in particular, Marissa Rollnick, Marié du Toit, and the students from the University of Venda. South Africa's beauty and hospitality will never be forgotten.

References

Adepetun, A. (2015, June 17). Africa's mobile phone penetration now 67%. *The Guardian*, http://guardian.ng/technology/africas-mobile-phone-penetration-now-67/

Bacca, J., Baldiris, S., Fabregat, R., Graf, S., & Kinshuk. (2014). Augmented reality trends in education: A systematic review of research and application. *Educational Technology & Society, 17*(4), 133–149.

Bujak, K. R., Radu, I., Catrambone, R., MacIntyre, B., Zheng, R., Zheng, G. G., & Golubski, G. (2013). A psychological perspective on augmented reality in the mathematics classroom. *Computers and Education, 68*, 536–544.

Chibale, K. (2015, July 2). Higher education in Africa: Our continent needs science, not aid. *The Guardian*. Retrieved from https://www.theguardian.com/global-development-professionals-network/2015/jul/02/higher-education-in-africa-science-not-aid

Chiu, M. H. (2012). Localization, regionalization, and globalization of chemistry education. *Australia Journal of Education in Chemistry, 72*, 23–29.

Chiu, M. H. (2015a). CNT at https://www.youtube.com/watch?v=HxMJc_-bK4c

Chiu, M. H. (2015b). *Chemistry education in Africa: Opportunities and challenges (plenary talk)*. Paper presented at ACRICE 2015---2nd African Research Conference of Chemical Education, November 22–27, University of Venda, Venda, South Africa.

Chiu, M. H., & Duit, R. (2011). Globalization: Science education from an international perspective. *Journal of Research in Science Teaching, 48*(6), 553–566.

Dolin, J., & Krogh, L. B. (2010). The relevance and consequences of PISA science in a Danish context. *International Journal of Science and Mathematics Education, 8*(3), 565–592.

du Toit, M. (2015, November). *Chemistry for the masses: The value of small scale chemistry to eliminate misconceptions and re-establish practical work in diverse communities*. Paper presented at the 2nd African Conference on Research in Chemical Education (ACRICE), University of Venda Venda, South Africa.

Duh, H. B. L., & Klopfer, E. (2013). Augmented reality learning: New learning paradigm in co-space. *Computers and Education, 68*, 534–535.

Gilbert, J., & Treagust, D. F. (2009). *Multiple representation in chemistry education*. Dordrecht: Springer Science+Business Media B.V.

Hodson, D. (2003). Time for action: Science education for an alternative future. *International Journal of Science Education, 25*(6), 645–670.

Ibáñez, M. B., Di Serio, A., Villarán, D., & Kloos, C. D. (2014). Experimenting with electromagnetism using augmented reality: Impact on flow student experience and educational effectiveness. *Computers & Education, 71*, 1–13.

Jensen, W. B. (1998). Logic, history, and the chemistry textbook I. Does chemistry have a logical structure? *Journal of Chemical Education, 75*(6), 679–687.

Johnstone, A. H. (1982). Macro-and micro chemistry. *School Science Review, 64*(227), 377–379.

Kozma, R. B., & Russell, J. (1997). Multimedia and understanding: Expert and novice responses to differentrepresentations of chemical phenomena. *Journal of Research in Science Teaching, 34*(9), 949–968.

Martin, M. O., Mullis, I. V. S., Gonzalez, E. J., & Chrostowski, S. J. (2004). *TIMSS 2003 international science report: Findings from IEA's trends in international mathematics and science study at the fourth and eighth grades*. Chestnut Hill: International Association for the Evaluation of Educational Achievement (IEA), TIMSS &PIRLS International Study Center, Boston College. http://timss.bc.edu/timss2003i/scienceD.html

Martin, M. O., Mullis, I. V. S., & Foy, P. (with Olson, J. F., Erberber, E., Preuschoff, C., & Galia, J.) (2008). *TIMSS 2007 international science report: Findings from IEA's trends in international mathematics and science study at the fourth and eighth grades*. Chestnut Hill: International Association for the Evaluation of Educational Achievement (IEA), TIMSS & PIRLS International Study Center, Boston College. http://timss.bc.edu/TIMSS2007/sciencereport.html

Martin, M. O., Mullis, I. V. S., Foy, P., & Stanco, G. M. (2012). *TIMSS 2011 international results in science*. Chestnut Hill: International Association for the Evaluation of Educational Achievement (IEA), TIMSS & PIRLS International Study Center, Boston College. http://timss.bc.edu/TIMSS2011/international-results-science.html

McElhaney, K. W., Chang, H. Y., Chiu, J. L., & Linn, M. C. (2015). Evidence for effective uses of dynamic visualisations in science curriculum materials. *Studies in Science Education, 51*(1), 49–85.

Neumann, K., Fischer, H. E., & Kauertz, A. (2010). From PISA to educational standards: The impact of large-scale assessments on science education in Germany. *International Journal of Science and Mathematics Education, 8*(3), 545–563.

NGSS Lead States. (2013). *Next generation science standards: For states, by states*. Washington, DC: The National Academies Press.

Ogunniyi, M. B. (1988). Adapting western science to traditional African culture. *International Journal of Science Education, 10*(1), 1–9.

Organisation for Economic Co-operation and Development (OECD) (2004). *Learning for tomorrow's world: First results from PISA 2003*. OECD Publishing. http://www.oecd-ilibrary.org/education/learning-for-tomorrow-s-world_9789264006416-en

Organisation for Economic Co-operation and Development (OECD) (2007). *PISA 2006, science competencies for tomorrow's world: Volume 1: Analysis*. OECD Publishing. http://www.oecd-ilibrary.org/education/pisa-2006_9789264040014-en

Organisation for Economic Co-operation and Development (OECD). (2008). *PISA 2006, Volume 2: Data*. OECD Publishing. http://www.oecd-ilibrary.org/education/pisa-2006_9789264040151-en

Organisation for Economic Co-operation and Development (OECD). (2010). *PISA 2009 results: What students know and can do. Student performance in reading, mathematics and science (Volume I)*. OECD Publishing. http://www.oecd-ilibrary.org/education/pisa-2009-results-what-students-know-and-can-do_9789264091450-en

Organisation for Economic Co-operation and Development (OECD). (2014). *PISA 2012 results: What students know and can do. Student performance in mathematics, reading and science (Volume I)*. OECD Publishing. http://www.oecd-ilibrary.org/education/pisa-2012-results-what-students-know-and-can-do-volume-i-revised-edition-february-2014_9789264208780-en

Poushter, J. (2016, February). *Smartphone ownership and Internet usage continues to climb in emerging economies.*Retrieved from http://www.pewglobal.org/2016/02/22/smartphone-ownership-and-internet-usage-continues-to-climb-in-emerging-economies/

Rollnick, M. (1990). African primary school teachers—What ideas do they hold on air and air pressure? *International Journal of Science Education, 12*(1), 101–113.

Rollnick, M., Bennett, J., Rhemtula, M., Dharsey, N., & Ndlovu, T. (2008). The place of subject matter knowledge in pedagogical content knowledge: A case study of South African teachers teaching the amount of substance and chemical equilibrium. *International Journal of Science Education, 30*(10), 1365–1387.

Taber, K. S. (2013). Revisiting the chemistry triplet: drawing upon the nature of chemical knowledge and the psychology of learning to inform chemistry education. *Chemical Education Research and Practice, 14*, 156–168.

Tahabalala, S. (2015, July 13). Africa's smartphone market is on the rise as affordable handsets spur growth, *Quartz Africa*. Retrieved from http://qz.com/451844/africas-smartphone-market-is-on-the-rise-as-affordable-handsets-spur-growth/.

Talanquer, V. (2011). Macro, submicro, and symbolic: The many faces of the chemistry "triplet". *International Journal of Science Education, 33*(2), 179–195.

United Nations. (2015). *The 2015human development report: Work for human development.* New York: Author http://www.undp.org/

Wu, H. K., Lee, S. W. Y., Chang, H. Y., & Liang, J. C. (2013). Current status, opportunities and challenges of augmented reality in education. *Computers & Education, 62*, 41–49.

Yore, L., Anderson, J. O., & Chiu, M. H. (2010). Moving PISA results into the policy arena: Perspectives on knowledge transfer for future considerations and preparations. *International Journal of Science and Mathematics Education, 8*(3), 593–609.

Chapter 2
Achieving the Aims of Practical Work with Microchemistry

John Bradley

1 Introduction

"Chemistry is fundamentally an experimental subject....education in chemistry must have an ineluctable experimental component." So states an IUPAC report (2000), expressing thus a core belief of chemical scientists and chemistry educators. It leaves open what more specific aims the experimental component of chemistry education might have and how they might be achieved. Although these issues have been answered in accepted ways for some decades, changes in the socio-economic context prompt reflection. Central to the contextual changes are achieving Education for All, population growth, and environmental sustainability.

A century ago, education was not for all. Science education in particular was offered to a minority of those to be educated. For some of this minority, laboratory practice was provided with the image of tertiary education and the research or post-graduate laboratory very much in mind. Providing for a minority only, meant that the cost of it was not of great concern. This elitist situation has long since gone in the primary and secondary school system and even to some extent at the tertiary level. Education for All has no limits as a concept, and this is now coupled with a rapid growth of the global population and the concomitant climate change. Nowhere are these major trends more evident than in Africa today. However, chemistry education in Africa has yet to respond significantly to these changes. Chemistry curricula remain largely those developed in previous decades to provide a "normal chemistry education" (van Berkel 2005). Clearly, they have validity for the minority who plan a career in chemistry, but not for all. In following through on the educational imperative to serve all, curricula need to be reconceptualised, with aims such as those suggested by Reid (2012):

J. Bradley (✉)
RADMASTE Microscience Project, School of Education, University of the Witwatersrand, Johannesburg, South Africa
e-mail: John.Bradley@wits.ac.za

L. Mammino, J. Apotheker (eds.), *Research in Chemistry Education*,
https://doi.org/10.1007/978-3-030-59882-2_2

(a) understanding something of the way the world works;
(b) appreciating the huge contribution of chemistry in human welfare;
(c) appreciating how chemistry gains its insights.

In a number of countries (for example USA, UK, Germany, Netherlands), new chemistry curricula that are in tune with these aims have emerged. They take account of the concept of scientific literacy, theories of motivation, and approaches of situated learning (Nentwig et al. 2007; Apotheker 2014). The slogan "Chemistry in Context" summarises to some extent the curriculum emphasis (Roberts 1982). Within Africa, it remains to uncover what "contexts" would be appropriate, but they will not be identical to those in Europe or N America.

Given a complete change of socio-economic context and, perhaps, of chemistry curricula, how then can the IUPAC report's imperative be addressed? If "normal chemistry education" is no longer regarded as appropriate for the majority, does this mean different practical activities with different aims? Can chemistry education escape from "normal practical work" (to use van Berkel's description) and adopt practical activities suited to a different curriculum emphasis? Looking at it realistically, the costly old ways, which were adopted for the minority, are not affordable for all. Furthermore, the environmental impact of all science students generating chemical wastes in the traditional quantities cannot be accepted. Indeed surely it is sending a wrong message to use 20 ml in a titration when 2 ml will suffice, or to use 5 ml of reagent when 2 drops will do equally well. For all such reasons the author has preferred for several years specially-designed microscale equipment, rather than the traditional scale equipment. This paper considers the author's and his group's experiences in Africa so far, and argues that pilot projects are required that have a strong research component.

2 The Aims of Practical Work in School Chemistry

Much has been written about the aims of practical work in a school curriculum. The succinct listing of Woolnough and Allsop (1985) is one appropriate to our needs:

1. Motivation
2. Developing practical skills
3. Learning the scientific approach
4. Gaining a better understanding of theoretical aspects of the subject.

Chemistry teachers are often unclear about the aims of practical work and may see an activity as a necessary evil required by a national curriculum. They may regret the time spent which, in their view, could have been better used in preparing learners for written examinations. Even those who have a less jaundiced view may simply think practical work is a good thing, without thinking how a particular activity might be designed to achieve one or more of the aims.

Such realities no doubt explain why the measurable benefits of practical work have generally been disappointing (Dillon 2008). By contrast learners generally enjoy practical work, suggesting that benefits are there to be realized if educators apply their minds to the challenge.

3 Problems of Practical Work in School Chemistry

Some problems have been mentioned above, but there are plenty more:

- aims often confused and unfocused
- insufficient time in the curriculum
- poor quality or inexperience of teachers
- inadequate or non-existent technical assistance
- cost of laboratory, its equipment and its maintenance
- safety and environmental regulations limit scope
- examination system assesses bookwork predominantly or exclusively.

These problems afflict practical work in the curriculum regardless of the scale (macro or micro) of the equipment used. National curriculum developers and teacher training institutions must take the blame for the continuing existence of some of these problems, but financial limitations are real and will not soon be ameliorated. These limitations directly affect points four and five in the above list, and within Africa particularly, they are severe. A chemistry teacher without resources for practical activities and technical assistance can quite reasonably claim to be disabled. Improvisation may be a noble calling, but it is not a reasonable way of life.

4 Ameliorating the Problems by Using Microchemistry

Microscale chemistry can be part of the solution, especially if low-cost plastics are the dominant material used in the equipment. Many useful components (such as droppers and microwell plates) exist as stock items, and these immediately open the door to worthwhile hands-on, learner activities at very low cost. The author and his group have greatly extended the capability of such equipment by using specially-designed additional components, such as the Comboplate and well lids, which permit the preparation and use of gases, electrochemistry, etc. The versatility is there to cater for creative teachers and learners, almost without limit. The limits are now really with the chemicals.

The chemicals limitation is again a serious one, but one that microchemistry ameliorates: the small scale of work means a lesser consumables cost burden. At the same time, it ameliorates safety and environmental concerns: smaller quantities mean greater safety and less waste disposal problems. Even the traditional requirement of a science laboratory falls away (however, the rules of correct use and handling of chemicals of course remain).

These points link naturally to the issues of time and convenience. It is true that school chemistry curricula are often overcrowded, and this is one reason why practical work is neglected. There is much evidence that practical activities conducted on microscale take place more quickly than the equivalent ones on traditional scale. Furthermore, with groups of learners looking after their own kits, teachers have less need for technical assistants and waste less time in administrative tasks. They still need to manage the activities, but can choose the most appropriate point in a regular classroom lesson to implement them. A well-designed and quick activity can leave enough time for classroom discussion of it within the one lesson.

The convenience for both teachers and learners, together with the quick and direct link between the domain of observables and the domain of ideas (Abrahams and Reiss 2010), help to motivate both parties engaged in the teaching and learning. Inquiry-based, student-centred instruction, as advocated by Lamba (2015), becomes accessible to many more classrooms than heretofore.

The economic, administrative and pedagogical advantages of microchemistry for school systems would seem to make an irresistible case for wider implementation (Bell et al. 2015).

5 Implementing Microchemistry in the School System

Over 20 years our group has sought to introduce the advantages of microchemistry to educators in different countries. Introductory workshops have been conducted (with the support of UNESCO, IUPAC, IOCD and others) in more than 80 different countries – about half of them in Africa. In the latter cases, practical work is certainly part of the curriculum, but it is rarely done. Above all this is because there are no funds available. The participants were generally selected by the national Ministry of Education and included education officials, school teachers, teacher trainers. They were all well aware of the gap between the curriculum ideals and the school realities, insofar as practical work is concerned. The majority of the participants expressed enthusiasm for the microscale approach and in some cases the Ministry took further steps, including the training of a group of selected teachers to prepare for a wider implementation. They saw the potential of the microscale approach to practical work as a feasible solution to this problem. Yet beyond this, rather little has usually developed (Bradley 2016), although there have been exceptions. One of these has been in Cameroon where, starting in 1998, an Inspector of Schools persuaded successive Ministers of Education to invest in microchemistry kits and in teacher training over a period of time. This commitment has attracted further support from UNESCO. Another recent project (2011–2014), again supported by UNESCO, saw the distribution of microscale science kits (chemistry, physics and biology) to 180 schools in Tanzania. This time an attempt was made to monitor and evaluate the project, and it seems clear from the draft report that considerable difficulties were experienced. Although some positive achievements were reported, the majority of schools experienced disappointment due to lack of teacher preparation and support.

These disappointments in Africa reflect similar disappointments in South Africa, where so much effort has been invested in developing a range of low-cost, microscale science equipment and producing them in the country. In the early years of the development, government and donors enthusiastically supported the introduction of microscience kits in schools. The gap between curriculum ideals and realities also existed in South Africa and these kits seemed to be the way to bridge this gap. Over time it became clear that teacher preparation and support for this development was quite inadequate. Whilst teachers continued to be enthusiastic when introduced to the microscale approach, they mostly failed to use the equipment in their teaching. Focused studies by researchers could show the learning benefits attributable to using the kits in a suitable context, but these could not be realized when replication was attempted more widely (Bradley 2016).

In summary, the experience of the past 20 years has been that microchemistry is welcomed as an idea, and has demonstrated success under controlled research conditions, but has largely failed to stimulate practical work when implemented more widely.

6 The Zone of Feasible Implementation (ZFI)

Our frustrations with the introduction of microchemistry reflect frustrations experienced with educational innovations generally. According to Rogan and Grayson (2003) the success and uptake of any educational innovation depends upon several factors, which define a zone of feasible implementation. They make a number of propositions for success in this respect, as follows:

1. Innovation should be just slightly ahead of existing practice.
2. Capacity to support innovation needs to be developed concurrently.
3. Outside support should not exceed the capacity of a school to use it.
4. All role players need an opportunity to reconceptualise intended changes in their own terms and for their own context.
5. Changing teaching and learning practices should be seen as a culture change, not just a technical change.
6. The ultimate aim must always be an improved learning experience.

In our case, the first proposition immediately signals a major difficulty: if as a teacher you have not ever implemented practical work in your classroom then doing so for the first time is more than "slightly ahead of existing practice"! Furthermore, continuing support and development strategies planned by Ministries of Education, were invariably weak. The assumption or pious hope, that teachers on their own could manage and reconceptualise their teaching and learning programmes, is demonstrably unsound. The fact is that implementing any science practical work in the majority of schools in Africa is a far bigger challenge than Ministries of Education have been prepared to admit.

Offering low-cost microscale practical science as a solution to the provision of school practical experiences has uncovered the familiar basic truth: it depends on the teachers.

In the short-term it will be the above-average teacher who will succeed; in the longer term the solution must lie in the preparation of new teachers. The new teachers should expect to undertake practical work and to have the equipment and chemicals when they start teaching.

In the light of this it may be concluded that pilot projects are required that have a strong research (monitoring and evaluation) component, in order to learn how the "ineluctable experimental component" can best be established in chemistry education in Africa.

7 Practical Work in the Context of Sustainable Development in Africa

Education for sustainable development (ESD) increasingly attracts debate around the World (Venkataraman 2009). Indeed, the first decade of ESD (2005–2014) initiated by the United Nations has recently been completed, and for chemistry, IUPAC has been active within this, as part of its engagement with the place of chemistry in sustainable development (Tarasova 2015). Practical work discussions must clearly be cognizant of policies and principles arising from these wider debates, and these discussions should be taking place in Africa at least as much as on any other continent.

The previous discussion of the advantages of microchemistry has mentioned the obvious reduced environmental impact of microscale chemistry practice. Chemistry educators should place emphasis on this as an important factor in choosing microchemistry rather than the traditional scale (Ibanez 2012). At the same time equity, in the education provided to the growing school population, is fostered as a result of the greatly-reduced costs of microscale work. These are significant advantages in the context of sustainable development on an increasingly populous continent.

But further curricular changes should be considered. If all learners are to be prepared for a sustainable future, the current adherence to "normal chemistry education" must be greatly diminished. Onwu and Kyle (2011) have argued that linking science education to issues of sustainable development could increase the perceived relevance of science for learners and thereby stimulate their interest in it. Whilst their case makes no reference to practical work, its relevance seems clear: ideas need to be linked with observables! It remains of course to design school chemistry curricula that have an appropriate emphasis. And in order for this to be a worthwhile exercise the principles of ZFI must not be forgotten. There does need to be a new direction with new aims such as Reid suggests, but full attention must be devoted to how the change of direction is to be achieved by classroom teachers (Ogunniyi 2011).

One example from South Africa will serve to remind us. A new curriculum in Physical Sciences for Grades 10–12 was introduced in 2005, the content of which was designed to be meaningful, accessible and relevant to all students (Ramnarain and Fortus 2013). New themes (Chemical Systems, Matter and Materials) appeared alongside more familiar-sounding ones (Chemical Change) and there was evident a shift of curriculum emphasis towards that of Chemistry in Context. Ramnarain and Fortus reported that both teachers and learners were positive about the new topics and saw their potential worth. However, teachers felt they lacked knowledge (including pedagogical content knowledge, PCK) to teach effectively. In consequence, they adopted teacher-centred strategies and avoided engagement with learners. Regrettably, the new curriculum was amended in 2011 to greatly reduce the curriculum time devoted to the new topics and to reinstate a more "normal chemistry education".

8 Conclusions

Achieving the aims of practical work, as suggested by Woolnough and Allsop (1985), deserves our continuing effort. Teachers are not against them; indeed they are for them. Microscale chemistry offers a realistic way of pursuing these efforts, but adopting this approach must be coupled with serious support and development of teachers. Above all in Africa, this is because teachers have generally never had significant opportunities to conduct chemistry classes in which hands-on practical activities by learners are a regular and central component. A culture change is therefore contemplated.

A further culture change should receive our attention: that is embracing a new curriculum emphasis that breaks with "normal chemistry education" (perhaps suited to a minority) and addresses the needs of the majority. Climate change is evident to many communities in Africa and education for sustainable development is a challenge that cannot be avoided. The low-cost, convenience and versatility of microscale chemistry will assuredly play a part in providing this education.

Abegaz (2016), then Executive Director of the African Academy of Sciences, provides a stimulating overview of challenges and opportunities for chemistry in Africa. He stresses:

> It is Africa's youth that will enable it to extricate itself out of poverty and take its place in the global fields of science, technology and innovation (STI). But this will not come easily. The other side of the coin is that this 'youth bulge' also makes investment crucial: to provide good quality, relevant education which will, in turn, lead to employment opportunities.

This perspective needs to be seized upon by Ministers of Education and Science and Technology, and by science educators at all levels.

References

Abegaz, B. (2016). Challenges and opportunities for chemistry in Africa. *Nature Chemistry, 8*, 518–522.

Abrahams, I., & Reiss, M. (2010). Effective practical work in primary science: The role of empathy. *School Science Review, 113*, 26–27.

Apotheker, J. (2014). The development of a new chemistry curriculum in the Netherlands: Introducing concept-context based education. *African Journal of Chemical Education, 4*(2), 44–63.

Bell, B., Bradley, J. D., & Steenberg, E. (2015). Chemistry education through microscale experiments. In J. Garcia-Martinez & E. Serrano-Torregrossa (Eds.), *Chemistry education: Best practices, opportunities and trends*. Weinheim: Wiley VCH.

Bradley, J. D. (2016). Achieving the aims of school practical work with microchemistry. *African Journal of Chemical Education, 6*(1), 2–16.

Dillon, J. (2008). *A review of the research on practical work in school science*. London: SCORE.

Ibanez, J. G. (2012). Miniaturizing chemistry: The ecological alternative. *African Journal of Chemical Education, 2*(1), 3–9.

IUPAC. (2000). Report of the education strategy development committee (p. 8).

Lamba, R. S. (2015). Enquiry-based student-centered instruction. In J. Garcia-Martinez & E. Serran-Torregrossa (Eds.), *Chemistry education: Best practices, opportunities and trends*. Weinheim: Wiley VCH.

Nentwig, P. M., Demuth, R., Parchmann, I., Gräsel, C., & Ralle, B. (2007). Chemie in Kontext: Situated learning in relevant contexts while systematically developing basic chemical concepts. *Journal of Chemical Education, 84*(9), 1439–1444.

Ogunniyi, M. B. (2011). The context of training teachers to implement a socially relevant science education in Africa. *African Journal of Research in Mathematics, Science and Technology Education, 15*(3), 98–121.

Onwu, G. O. M., & Kyle, W. C. (2011). Increasing the socio-cultural relevance of science education for sustainable development. *African Journal of Research in Mathematics, Science and Technology Education, 15*(3), 5–26.

Ramnarain, U., & Fortus, D. (2013). South African physical sciences teachers' perceptions of new content in a revised curriculum. *South African Journal of Education, 33*(1), 1–15.

Reid, N. (2012). Successful chemistry education. In *Proceedings 22nd ICCE-11th ECRICE,* Rome (pp. 290–297).

Roberts, D. A. (1982). Developing the concept of "curriculum emphases" in science education. *Science Education, 66*, 243–260.

Rogan, J. M., & Grayson, D. J. (2003). Towards a theory of curriculum implementation with particular reference to science education in developing countries. *International Journal of Science Education, 25*, 1171–1204.

Tarasova, N. (2015). Chemistry: Meeting the World's needs? *Chemistry International, 37*(1), 4–7.

van Berkel, B. (2005). *The structure of current school chemistry – A quest for conditions for escape.* Utrecht: CD-β Press, Centrum voor Didactiek van Wiskunde en Natuurwetenschappen, Universiteit Utrecht.

Venkataraman, B. (2009). Education for sustainable development. *Environment, 51*(2), 8–10.

Woolnough, B., & Allsop, T. (1985). *Practical work in science.* Cambridge: Cambridge University Press.

Chapter 3
Chemistry for the Masses: The Value of Small Scale Chemistry to Address Misconceptions and Re-establish Practical Work in Diverse Communities

Marié H. du Toit

1 Introduction

Why is chemistry so essential? Chemistry is the central science, and it is important in many different knowledge fields, including medicine, education, pharmacy, engineering, physics, nutrition sciences, forensics, materials sciences, polymers sciences, preparation, and many others. However, according to research and experience, school learners (Arends et al. 2015; Beaton 1996; Martin et al. 2004, 2012; Spaull 2013) and first year students (Potgieter and Davidowitz 2011) perform below par in chemistry. Learners, and in some cases teachers, have inadequate knowledge of the fundamental principles which underpin the study of chemistry. Compounding the problem of chemistry education is a serious shortage of skilled teachers in mathematics and science in South-African schools (Arends et al. 2015; Beaton 1996; Martin et al. 2004, 2012). The lack of skilled teachers undermines effective teaching. Ineffective teaching leads to the absence of "links" between existing knowledge and new knowledge and leaves memorization as only option (Marais and Mji 2009). Incomplete understanding of concepts leads to lack of self-discipline in learners and students to complete self-study and homework assignments. This vicious circle of under-qualified teachers and inadequate school systems and school resources results in poorly prepared learners and students, and this impacts negatively on the motivations and career-drive of learners and students (Marais and Mji 2009).

Furthermore, the complexity of chemistry as subject also contributes towards learners and students' poor performance. The abstractness and the language of chemistry make it difficult for students to understand and master the subject. Consequently, students experience chemistry as difficult and lose concentration and motivation to study it. Additionally, due to the challenging nature of the subject,

M. H. du Toit (✉)
School of Physical and Chemical Sciences, North-West University,
Potchefstroom, South Africa
e-mail: dutoitmarie.mh@gmail.com

L. Mammino, J. Apotheker (eds.), *Research in Chemistry Education*,
https://doi.org/10.1007/978-3-030-59882-2_3

learners and students do not receive encouragement and help from parents to master chemistry (Nbina 2012). Traditionally, chemistry comprises theoretical as well as practical work. Moreover, practical work is an essential part of science education (Millar and Abrahams 2009). In the words of Julia Buckingham, "Practical work is an integral part of science, it is not an add-on. It is something that encourages students to question, to explore – it excites them" (Adams 2014). Nevertheless, despite the importance of practical work, most chemistry teaching has deteriorated into lectures and memorizations to address time-constraints and curriculum overload. Some of the main reasons listed for why South African schools are not doing enough practical work are the lack of resources (many schools have resource barriers (Marais and Mji 2009)) and laboratories, time constraints, large classroom size and assessment pressures (Heeralal 2014). Moreover, school managements have difficulties with the notion that the physical sciences need more financial investments than other subjects. In addition, personal experience and literature show that teachers are discouraged from doing practical work because of time constraints, lack of self-confidence and motivation, lack of organization, fear of working with chemicals, anxiety about performing experiments, and insufficient understanding of relevant chemistry concepts (Kibirige et al. 2014). Compounding these problems is the remoteness of many rural schools, as is also the case in South Africa. Some do not even have adequate access roads. If practical work is to flourish and become the central activity of teaching chemistry, these issues need to be addressed.

The poor performance of learners and students in chemistry, combined with the awareness of the resource constraints of many schools, were the main driving force for the development of the commercially available MYLAB Small Scale Chemistry and Natural Science kits (Grade 4 to 12 (www.mylab.co.za). The value of small scale chemistry to increase learners' and students' understanding of chemistry concepts was researched in Ethiopia with positive results (Tesfamariam et al. 2014). Moreover, the international drive towards green and microscale chemistry substantiate the value of small scale chemistry (Tantayanon 2005).

The MYLAB kits include all the apparatus the learners need, all the chemicals, adequate and challenging worksheets based on school curriculum outcomes, memorandums for worksheets and preparation material (in the DVDs). No laboratories are necessary, nor electricity or running water. These small-scale chemistry kits are an easy and cost-effective way to reach large numbers of schools and pupils in rural areas (Du Toit 2012a) and provide a solution for easy transfer of knowledge and practical skills. The main advantages of using the kits include easy storage, ease of use, less expensive than standard laboratory equipment if there are breakages, and the use of small amounts of chemicals. Furthermore, experiments can be done easily, seen clearly, performed in shorter time and are safer. Also, the wide range of experiments that can be performed makes the kits especially valuable. Short, quick, easy experimentations with small scale chemistry kits can be used in ordinary classrooms to highlight, discover, test and investigate chemistry concepts. Thus, the independence of the chemistry kits of classroom facilities, electricity, and running water, make them imminently suitable for diverse communities and under-resourced schools. Moreover, onsite training workshops are offered for schools purchasing these kits. Small scale chemistry workshops try to address the main concerns which teachers experience, such as the lack of self-confidence, adequate background knowledge, sufficient theoretical knowledge, practical skills, and knowledge about chemicals and safety data (Du Toit 2006, 2014, 2015, 2016).

However, when doing practical work, one needs to simultaneously address trouble-shooting (what to do if an experiment does not work or give the desired results) and misconceptions (wrong chemistry concepts held by teachers or learners). Misconceptions are notoriously resistant to conceptual change; therefore, teachers (and learners) need to be confronted with the correct concept through their own investigation (Hewson 1992; Taber 2002). Various strategies can be followed to bring about conceptual change (Hewson 1992; Kyle and Shymansky 1989; Riordan 2012). For example, a simple four-step approach (Bybee et al. 2008), also applied in this study, involves the following components:

1. first the action then the words;
2. talk through the new concept;
3. teach the concept to someone else;
4. do not let the concept die;

To apply these action-steps in chemistry teaching, practical work is of vital importance. There are many topics that learners and students do not fully understand, and the school systems do not provide enough time for practice and reinforcement of concepts (Ali 2012; Nbina 2012). However, small scale chemistry experimentation offers one of the solutions, especially useful in implementing the first step of teaching for conceptual change: first the action then the words. It makes it easy to confront learners and students with the correct scientific concepts (Niaz 1995). With the small-scale chemistry kits, learners and students can discover the concepts themselves, which is a precursor to better and deeper understanding by involving more than one of their senses. Physical confrontation and own explanations of chemical phenomena contributes to active learning, with improved long term effects. The focus of this study is to show that the use of small-scale chemistry kits, complemented by onsite workshops, is valuable to allow access to chemistry to all, especially reaching remote schools and improving chemistry subject knowledge by re-establishing practical work in schools. Through the evaluation of 22 years of onsite workshops, the author specifically wants to illustrate: (1) the effectiveness of MYLAB sets as a replacement for an entire laboratory; (2) how small-scale experiments can be used to address and improve misconceptions regarding chemical reaction types ; and (3) the value of onsite workshops to train chemistry teachers and reach remote communities.

2 Methodology

In order to help re-establish practical work and address misconceptions in chemistry, numerous workshops have been held over South Africa as part of the MYLAB project. The MYLAB project has been running for 22 years, totalling 260 workshops for 1950 teachers from 1800 different schools across the country (see Fig. 3.1 and Appendix A), as well as a large number of learners.

2.1 *Onsite Workshops*

Approximately 260 onsite workshops were held to re-establish practical work in diverse communities and especially rural communities in South Africa (Du Toit 2006, 2014, 2015, 2016). Workshops were also held on the North West University campus for pre-service and in-service teachers (Du Toit 2012b). Venues for workshops included science centres, schools and community halls. Due to financial constraints, workshops were done on demand: a provincial education department, a school or a financial institution requested a workshop or workshops. For the same financial constraints reasons, follow-up workshops were not always possible.

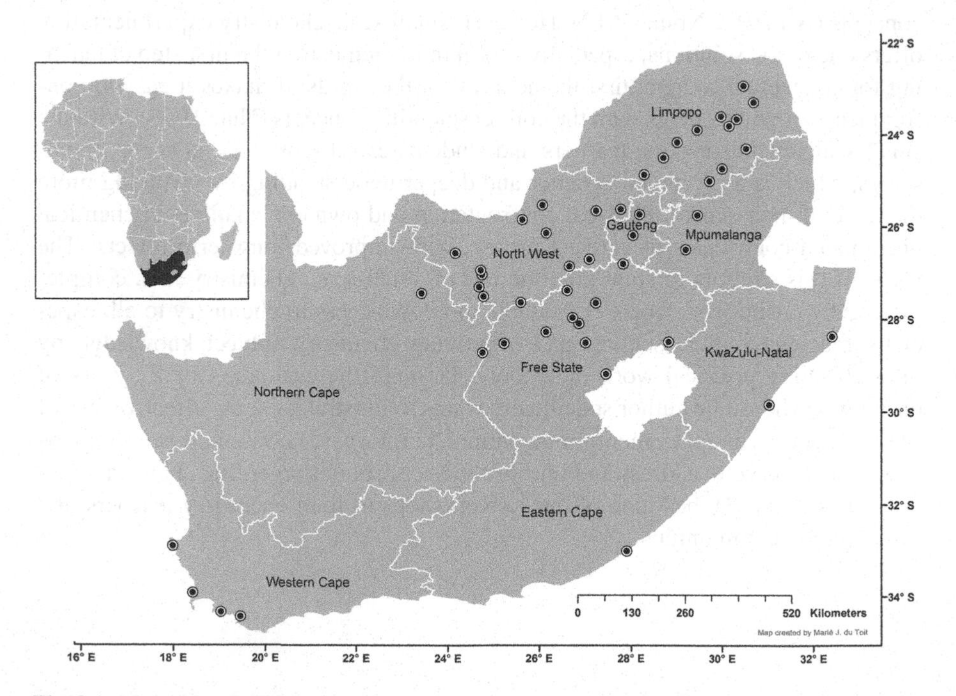

Fig. 3.1 Map of South Africa indicating the location of the towns and villages in which workshops where conducted. The inset map indicates the position of South Africa within the African continent. Appendix A gives the names and positions of the towns

The participant samples were random and consisted mostly of the science teachers and/or subject advisors of the school requiring the workshop, or of schools in the region. The maximum number of teachers in a workshop group was 25 (although larger numbers have been accommodated), to give high quality individual attention to each participant. The workshop presenter was supported by one to three workshop facilitators. Teachers were usually expected to do individual work within the workshops, because in their classrooms they are on their own. When groups were larger than 25, the teachers were allowed to work in groups to solve the experimental problems. The aim of this arrangement was to give attention to as many teachers as quickly as possible, to prevent individual frustration when problems arose during training.

The workshops followed a set program and four to six experiments were performed daily according to the South African school curriculum, to support the teachers in their teaching endeavours. Some workshops were conducted **formally**, where teachers wrote tests and completed experimental worksheets as part of their formal assessments (Du Toit 2006; Tholo et al. 2006), and other workshops were **informal** with no formal assessment (Du Toit 2014, 2015, 2016). Formal workshops involved separate sessions for different grade groups (grades 10, 11 and 12), and the teachers had to prepare for a preliminary test on the theory underlying the practicals corresponding to one of the groups. On the other hand, in informal workshops, teachers of different grade groups (such as grades 10 to 12, or grades 8 to 9, or grades 4 to 7) were teamed together, and concentrated attention was given to difficult concepts and practical experiments for each specific grade group. No formal written work was done and no assessment was conducted. However, teachers received printed hand-outs and made personal notes. The atmosphere in informal workshops was relaxed and discussion was encouraged. The results of informal workshops are qualitative and not quantitative.

Formal and informal workshops play an important role in teaching chemistry for the masses. Formal workshop programs are used when there is ample time and it is possible to work towards deeper understanding and a more critical evaluation of what is being done, be it experimental investigation or proving or disproving misconceptions. Informal workshops are used when there are time constraints and one has to do as much as possible in the time available.

2.2 Materials

The utilised teaching materials are the MYLAB small-scale chemistry (SSC) kits, as shown in Fig. 3.2. The toolbox of the kit is 42 cm × 20 cm × 23 cm and contains the complete SSC laboratory. The size of the 5ml glass test tube determines the size of the rest of the apparatus. The top tray contains all the chemicals needed for grades 10 to 12 chemistry. The middle tray contains the bigger apparatus pieces such as the back plate, base plate and test tubes (see Fig. 3.2). The bottom tray contains the smaller apparatus pieces like the stoppers, spatula, test tube brush and electrodes. The support material comprises worksheet manuals for learners and for teachers and one DVD per grade.

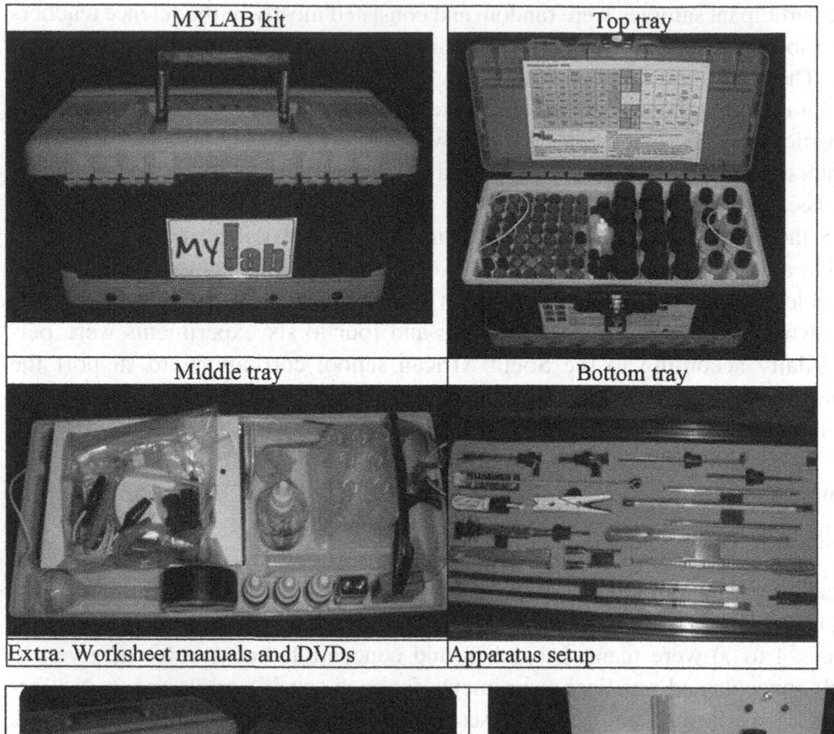

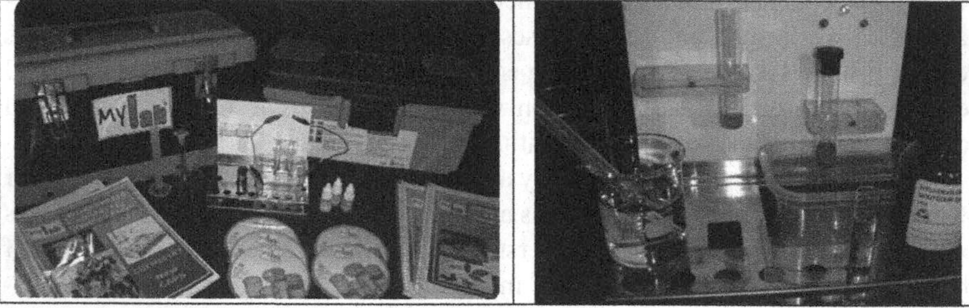

Fig. 3.2 The MYLAB kit and some of its components and setups

The MYLAB kits are tailor-made for hands-on teaching of natural science and chemistry at school level in South Africa. They provide an inexpensive solution to the problem of inadequate apparatus at school. They address the problem of under-achievement in chemistry and natural sciences by making practical work easy, cost-effective, safe, and quick. They are designed so that students can learn through self-experimentation. They are portable and versatile. They are complete and no laboratory is needed. What makes the MYLAB kit unique is that it is a compromise between the conventional and micro-scale apparatus: it is a small scale laboratory with apparatus of durable materials. According to green chemistry principles, small amounts of chemicals are used. All Chemistry experiments from grade 10 to 12 and all natural science experiments from grade 4 to 9 can be performed with the MYLAB kits.

2.3 Experiments and Misconceptions

Numerous misconceptions have been identified for chemistry. It is preferable to choose misconceptions that arise in class, during practical work or in workshops, while one is addressing or explaining a certain topic. Teachers need to keep the strategy for conceptual change in mind and address misconceptions accordingly (Bybee et al. 2008). A large number of conceptual change strategies are available in literature, all with the same aim (Hewson 1992; Kyle and Shymansky 1989; Riordan 2012), and anyone of them can be used. According to Hewson (1992), the ultimate goal is to teach learners and students two things: (i) "to form the habit of challenging your concept with other concepts and (ii) to develop appropriate strategies for having alternative conceptions compete with one another for acceptance." A number of misconceptions regarding general chemistry are reported in literature; Table 3.1 recalls a subset from (Horton 2007).

In formal workshops focusing on misconceptions, students (in-service teachers and pre-service teachers) were given specific examples of documented misconceptions, like those reported in Table 3.1, and were asked to design a strategy to correct them, by answering the following questions:

- What is the misconception?
- What is the correct concept?
- What is the possible origin of the misconception?
- How can the concept be corrected? (Teaching for conceptual change.)

Table 3.1 Selected misconceptions regarding general chemistry (Horton 2007)

#	Misconceptions
1	"Chemical reactions are caused by mixing of substances"
2	"Chemical reactions are reactions which produce irreversible change"
3	"The original substance vanishes "completely and forever" in a chemical reaction."
4	"Chemical reactions between gases are simply mixing".
5	"Physical changes are reversible while chemical changes are not".
6	"Precipitation reaction results in change in mass".
7	"Mass increases because a solid weighs more than a liquid"
8	"Mass is lost in combustion".
9	"Testing for acids can only be done by trying to eat something away".
10	"When Mg is placed in aqueous HCl, the acid is the driving force, because it (the acid) is very strong."
11	"Mixing an acid with a base (without regard to quantities) neutralizes the base resulting in a neutral solution."
12	"In neutralization all the H and OH ions are cancelled."
13	"If a reaction doesn't involve oxygen it is not oxidation."
14	"Reduction is the removal of oxygen in a reaction."
15	"Electrons can flow through aqueous solutions without assistance from the ions."

They were also asked to submit a report with their results and conclusions. The cases shown in Table 3.2 are examples of how teachers can teach for conceptual change in addressing misconceptions.

Teachers were supplied with a MYLAB kit each and they had to show the conflict between the misconception and the correct concept. They had to illustrate the correct concept by doing an experiment and to explain the concept to a peer. The workshop presenter facilitated the process. In the workshop report, the teachers had to answer the four questions convincingly. Correcting misconceptions is a very important part of teaching and thus teachers should receive appropriate training in how to perform conceptual change in practice (Niaz 1995; Riordan 2012).

No formal written reports are required in informal workshops. The main drive is discussion, argumentation, teamwork and even debate. The participants (or groups of participants) of the workshop receive the misconception as a statement and they have to analyse, discuss and correct the misconception by doing an experiment. Individual teachers prepare informal report, and each teacher or each group presents an oral summary of their results. Conceptual change strategies are used to address a subset of the documented misconceptions (Box 3.1) and are implemented in the form of practical activities using the MYLAB kits.

At one of the informal workshops, for example, 48 participants (in-service teachers) were divided in eight teams of six teachers each. Each team had the instructions listed in Box 3.1. They could choose any misconception and answer the questions, write a short report and present the report to all the workshop participants as an audience. Each team was provided with a facilitator (one of the workshop presenters or a knowledgeable person or a subject specialist).

3 Results

3.1 Results of Formal Workshops

The results are discussed in the form of case studies. Each box gives an example of a misconception or misconceptions and related questions. For instance, in example 1, shown in Box 3.2, the misconception M1 is given with four sub-questions (Q1 to Q4) that the student-teachers must answer and discuss.

All four questions in Box 3.2 were seen as challenging and difficult to answer. In a group of 24 fourth year pre-service teachers (student-teachers) in a 2012 workshop, the following results (see also Appendix B) were obtained. The average mark for the complete question was 30.8%. The average mark for Q1 was 29%. Seven student-teachers identified the chemical reaction type correctly. Ten student-teachers decided it was an acid-base reaction and some added that it was ion-exchange. One student-teacher decided the reaction was a substitution reaction. Three student-teachers identified the reaction as a gas-forming reaction. One mentioned that the gas-forming reaction was also an ion-exchange. Two student-teachers decided on

Table 3.2 Examples of analysis of three misconceptions selected from Table 3.1, including examples of experiments that can be used to address them and bring about conceptual change

Misconception 1	"Chemical reactions are caused by mixing of substances" (Horton 2007)
Origin	The use of language and macroscopic observation may be the possible origins of this misconception, e.g. Na and water or an acid and a base are "mixed together" according to learners. Stoichiometry, the law of definite proportions and reaction conditions are completely ignored. The difference between mixtures and chemical reactions are not properly defined and explained.
Correct concept	The above-mentioned incorrect statement can be corrected by stating that "Chemical reactions take place when substances react with each other".
Teaching for conceptual change	Mix iron filings and sulfur powder. Mixing gives a mixture Fe and S. The mixture can be separated by physical means (magnet). NO chemical reaction takes place. $Fe + S \rightarrow FeS$ A chemical reaction takes place when the two reactants are heated together and then produce FeS. A different substance with different properties is formed and the original reactants can only be obtained with separation by chemical means. The chemical substances, iron and sulfur, react in a fixed ratio of 1:1.
Misconception 2	"Chemical reactions are reactions which produce irreversible change" (Horton 2007)
Origin	Language and macroscopic observation. Students observe magnesium burning in oxygen and there is no way they can recover the magnesium. In the same way Na reacts with water.
Correct concept	All chemical reactions are reactions which produce reversible change. (Indicated by equilibrium constant.)
Teaching for conceptual change	Cobalt chloride and water $\left[Co(H_2O)_6\right]^{2+}(aq) + 4Cl^-(aq) \rightleftharpoons [CoCl_4]2^-(aq) + 6H_2O(l)$ pink blue Cobalt chloride dissolved in ethanol is blue. If water is added it turns pink. If hydrochloric acid is added it turns blue again. The equilibrium can be moved backwards and forwards.
Misconception 10	"When Mg is placed in aqueous HCl, the acid is the driving force, because it (the acid) is very strong." (Horton 2007)
Origin	Students know that HCl is a strong acid (according to theory) and decide the strong acid is the driving force. Students lack scientific understanding of the concept of the driving force of a reaction (the chemical phenomenon).
Correct concept	When Mg is placed in aqueous HCl, the electron transfer is the driving force of the reaction. Electron transfer will be the driving force of the reaction between an acid and a metal whether the acid is strong or weak.
Teaching for conceptual change	React magnesium with different acids (strong and weak) and observe what happens. Both strong and weak acids react with magnesium thus the argument about the strength of the acid, as reason for the driving force, is not valid. If the strength of the acid is not the driving force what is the driving force? Careful explanation to proceed from macroscopic observation, to sub-microscopic theory and symbolic representation is necessary. Half reactions of oxidation and reduction to the complete redox reaction need to be understood by students.

Box 3.1: A Typical Example of a Misconception Assignment at an Informal Workshop

The Example is from a Workshop at the University of Venda in November 2015 (Du Toit 2015)

How to address misconceptions

1. What is the scientifically incorrect conception?
2. What is the origin of the incorrect conception?
3. What is the correct scientific conception?
4. How can the teacher teach with practical experiments to change the alternative concept to the correct scientific concept? (teaching for conceptual change)

Misconceptions

1. Solute (salt, sugar) disappears when dissolved.
2. Weightless matter can exist.
3. Substances prepared in different ways cannot be the same substance; the way of preparing a substance is one of its properties.
4. Water disappears as it evaporates
5. Air weighs less when it is expanded.
6. Less dense means weighs less
7. A kilogram of lead weighs more than a kilogram of water.
8. The weight of a substances changes when it changes phase.
9. The products of chemical reactions need not have the same mass as the reactants.
10. Objects float because they are light.
11. A balloon is lighter if air is blown into it.
12. A small paper clip floats better than a large paper clip.
13. Hot and cold are different kinds of substances.
14. Drops of water on the outside of a cold bottle come from inside the bottle.
15. Drops of water on the outside of a cold bottle are from hydrogen or oxygen combining.
16. Bubbles mean boiling.
17. Melting and dissolving is the same thing.
18. Sugar (or salt) dissolving in water is a chemical change.
19. A rusting nail will lose weight.
20. In electrolytic cells, water is unreactive towards oxidation and reduction.
21. No reaction will occur if inert electrodes are used
22. Endothermic reactions cannot be spontaneous.

Box 3.2: Example 1: A Misconception Proposed in a Formal Practical Workshop in 2012

The abbreviations (M1 for the misconception, E1 for the example, Q1, Q2, Q3, Q4 for the four sub-questions asked) are used in Appendix B to indicate the quantitative scores student-teachers achieved in their workshop reports.

Misconception 1 (M1) of example 1 (E1):

"When Mg is placed in aqueous HCl, the acid is the driving force, because it is very strong."

Questions:

Q1 What type of chemical reaction takes place between Mg and HCl?
Q2 What is the driving force of the reaction?
Q3 What is the specific misconception in the given statement?
Q4 How is the teacher going to teach for conceptual change to correct this misconception?

the reaction as a precipitation reaction and one indicated that the precipitation reaction was also a redox reaction. One student-teacher identified the reaction as an exothermic reaction. The average mark for Q2 was 40%. The chemical phenomena responsible for the reaction were given as electron transfer (9 student-teachers; the correct answer), proton transfer (7 student-teachers), formation of an insoluble gas (3 student-teachers), ion-exchange (3 student-teachers), the formation of an insoluble compound (1 student-teacher) and the rearrangement of atoms (1 student-teacher). Two student-teachers considered acid-base reactions as the reaction type, but electron transfer as the driving force of the reaction. One student-teacher gave both electron transfer and proton transfer as the driving forces for an acid-base reaction. The average for identifying the specific misconception in the statement was 36.5%. The average for how the teacher is going to teach for conceptual change was 28.6%.

Misconception 2 (M2) of example 1, with related questions similar to those in Box 3.2, compared covalent bonds and intermolecular forces through the statement: "*The strength of covalent bonds and intermolecular forces are similar*". Out of 24 student-teachers, three could neither identify the incorrect concept nor the correct concept, eight identified the misconception correctly; 16 could say that covalent bonds are stronger than intermolecular forces, but not say why they believed that the bonds are stronger. Five student-teachers could correctly define covalent bonds and intermolecular forces, but could not formulate the argument to prove the statement wrong. One student-teacher stated that more energy is needed to break covalent bonds, but not why that is her conception. Two student-teachers used the distance between atoms for covalent bonds and the distance between molecules in intermolecular forces as grounds for their argument to refute the statement. None of the student-teachers could propose a practical experiment to help learners to comprehend the correct concept instead of the incorrect statement given.

Box 3.3: Misconception 1 (M1) of Example 2 (E2) and Related Question (Q1)

Learners have the following misconception (M1):

"Oxidation is the addition of oxygen in a reaction"

How is the teacher going to teach for conceptual change to correct this misconception?

Example 2 consisted of five different misconceptions; only the first 1 is considered here and shown in Box 3.3. The questions were asked in a formal practical workshop (2011) for 24 pre-service teachers (student-teachers). Their achievement scores are reported in Appendix C and clearly show that the question about conceptual change gave poor results. Similar results were obtained for the other four misconceptions, thus indicating the same inefficiency in using conceptual change strategies to address specific chemistry misconceptions.

The average mark for the question in Box 3.3 in a 2011 workshop was 42.7% (see also Appendix C). The student-teachers could give the generic answer about the four steps to be followed for conceptual change, but they could not describe the specific experiment they were going to use to confront the concepts of the learners and what specific explanations, assignments, tasks, they were going to give to facilitate the change. They did not know how to teach for conceptual change, similarly to what was found by Hewson (1992). Only 4 of the fourth-year student-teachers could indicate correctly that a redox reaction comprises two half reactions (one half reaction is oxidation and the other half reaction is reduction). Ten other misconceptions were identified in the answers of the 24 student-teachers. Five misconceptions were about writing chemical reaction equations and considering only oxidation without reduction. One student-teacher talked about reduction of the oxidation number instead of a chemical substance (ion, element, molecule or compound) and oxidation of the oxidation number. One student-teacher said oxidation and reduction is the same as ion-exchange. One student-teacher considered oxidation not only as addition of oxygen in a reaction, but also of addition of other elements. One student-teacher had the common mistake of "oxidation is the gain of electrons" and "reduction is the loss of electrons". Eight student-teachers completely ignored reduction and only considered oxidation.

In example 3 (Box 3.4), five misconceptions were listed and the same question was asked about each misconception, i.e., workshop participants were only asked to determine the origin of the mentioned misconceptions. The achievement scores of the 24 student-teachers participants are indicated in Appendix C.

The average mark for the question in Box 3.4 was 42.1% for the 24 participants of the workshop. The marks for M1 to M5 were respectively 29%, 50%, 76%, 29% and 26%. Student-teachers have little or no concept of phase changes (M1). 21 out of 24 participants thought the different physical states of matter in a chemical reaction equation indicate a phase change. Three of these 21 could correctly argue their

Box 3.4: Example 3: A Question About the Origin of Five Misconceptions Concerning Chemical Reactions

Learners have the following five misconceptions (M1, M2, M3, M4, M5):

M1	Chemical reactions are phase changes
M2	The original substance vanishes completely and forever in a chemical reaction.
M3	The H_2 bonds are not broken in forming H_2O
M4	Chemical reactions are caused by mixing of substances.
M5	Heat supplied or absorbed is the driving force in a burning candle.

What is the ORIGIN of the above-mentioned misconceptions?

way out of the misconception. The correct understanding of physical change and chemical change is also problematic. Chemical reaction, phase change, chemical bonding, chemical change and physical change are misunderstood (misinterpreted). Student-teachers used incorrect arguments to indicate the origin of the misconception because they themselves had this specific misconception (M1).

The results for M2 were better. Student-teachers mentioned the conservation of mass and reversible chemical reactions in their arguments to disprove the misconception of "vanishes completely and forever". Five student-teachers used the example of NaCl in water as their explanation and revealed the misconception that dissolution and evaporation are chemical reactions.

M3 was answered with few mistakes. Three student-teachers had three different misconceptions: one stated that, because H_2 is a gas, there is no bond to see; another stated that learners use the chemical equation $H_2 + O \leftrightarrows H_2O$ but when teachers use the crystal lattice as well to explain the misconceptions will be fewer; and the last one stated that bonds are broken because every H sits on a different orbital.

For M4, student-teachers and learners thought macroscopically, and thus believed the evidence of their eyes: the reactants are mixed, and a chemical reaction takes place. No sub-microscopic explanations were given. The difference between mixtures and chemical compounds were not used in the arguments. Only one student-teacher talked about chemical reactions and reactants reacting in the correct ratio to form products. Seven student-teachers had random misconceptions. Of these, student-teacher #1 saw the mixing of coffee powder and water as a chemical reaction. Student-teacher #2 stated that not all reactants mixed together react with each other, and that a force is needed to start the reaction; he also mentioned that reactions can start spontaneously without mixing. Student-teacher #3 stated that a reaction is the result of a mixture of two different substances, because the learners believed "one substance has to lose electrons onto the other substance". Student-teacher #4 stated that learners observe the mixing and do not know about "electron structures" that must be filled to become a noble gas, and that "the learners also do not know about activation energy and the influence of the environment".

Student-teacher #5 stated that learners "don't understand that a chemical reaction occurs when particles collide according to the collision theory until the substances have minimum kinetic energy which is the activation energy". Student-teacher #6 had the previously mentioned misconception that if salt and water are mixed, there is a chemical reaction. According to student-teacher #7, not enough variety in examples of reactions is responsible for the misconception about mixing, but this student's suggestion about using reactions with oxygen and combustion to prevent the misconception have no clear reasoning.

M5 highlighted the presence of a high number of misconceptions (listed in Box 3.5), and received the lowest marks (26%). One out of 24 student-teachers gave the correct argument. Three student-teachers did not even attempt to answer the

Box 3.5: The Misconceptions of Student-Teachers About a Burning Candle (Misconception M5 in Box 3.4)

- The driving force of the reaction is when the wick of the candle reacts with oxygen and an open flame.
- Some reactions have too much energy and liberate the energy in the form of heat, light, sound and so forth.
- Activation energy supplied or absorbed is the driving force in a burning candle.
- Learners don't understand the concept of the combustion of oxygen, because a candle liberates heat and light.
- The power is not the heat that is supplied, but the reaction between oxygen in the surroundings and the candle. The heat is caused by the gas or the reaction and it just starts the reaction.
- Learners associate the flame with heat and thus believe that the heat keeps the flame burning. In truth oxygen is the driving force of the combustion and heat is the result. Energy is released in the form of heat.
- Learners don't understand the concept of transfer of energy. They don't understand that the wax of the candle serves as the energy source. The chemical reaction is the driving force and the heat only serves to accelerate the process.
- Because the candle is ignited by a match that provides the heat, the learners assume that the heat provided to the candle lets the candle burn. They then don't know that oxygen really is the driving force that keeps the candle burning. The match is equal to the activation energy. They think that because the candle releases heat it keeps on burning, because they can feel the heat but not the oxygen.
- Heat absorbed is driven by the energy required to make the reaction take place so that is activation energy needed for reaction.
- Learners see that a match that burns is the cause of the candle burning and that heat is released. The candle was never put into a closed system so that the oxygen could become depleted to enable the learners to see that heat is just the activation of the reaction between oxygen and the candle wax and that oxygen is the driving force behind the burning candle.

question. Ten student-teachers had misconceptions. Two student-teachers gave meaningless answers. Eight student-teachers believed learners think macroscopically: "when the flame is there the candle burns, when the flame is gone, the candle is out so the flame must be the driving force for the burning candle".

3.2 *Results of Informal Workshops*

A workshop held at the University of Venda in November 2015 and involving 48 in-service teachers from schools in the neighbouring area (a rural area) is here considered as an example of informal workshop; it can be considered representative of other informal workshops that were held in different parts of South Africa. The Teachers were given the misconceptions in Box 3.1 and were asked to use conceptual change strategies to demonstrate the correct concepts. The teachers were enthusiastic and participated eagerly in trying to find a solution to the assignment, They chose the simplest or easiest misconceptions, but they could not progress in their assignment without guidance from the facilitators, due to lack of chemistry background knowledge and scientific creativity. However, the teachers worked well in groups and lively discussions resulted. Colleagues helped each other to clarify misconceptions and to devise possible experiments to address the misconceptions. The informal workshops were very interactive and cooperative problem-solving took place.

Each group had to assign different tasks to different people to achieve optimum results. Groups that assigned the roles of manager, strategist, recorder, experimenter and presenter quickly, and correctly for their specific team, were the most successful, whereas groups with too many chiefs did not prosper. The emphasis was on the process and not so much on the presentations at the end. Teachers needed practice in teaching for conceptual change, but the teachers were motivated, and the activities were appreciated. Several misconceptions were changed to correct concepts with the help of colleagues, workshop facilitators and workshop presenters.

3.3 *Some Excerpts from Post-Workshop Evaluation Questionnaires*

Approximately 260 workshops (both formal and informal) have been held. The feedback from the participants is overwhelmingly positive. Some comments are reported (in the participants' own words) in Box 3.6 (Du Toit 2006, 2014, 2015, 2016).

Box 3.6: Some Feedback and Comments from Participants of the More Than 260 Workshops Conducted

- I think this workshop is the beginning of a bright future to our learners. I will be improving in the next workshop. It was nice being here. I enjoyed it!
- This was fantastic. I wish this can be done (educators be workshopped) until being confident when teaching learners.
- If more of this work can be done quarterly, our results can improve drastically.
- Need more workshops to produce A's at matric level.
- Keep on shaping the nation by conducting the workshops, so as to build the future of our children, our country and the world as well.
- I believe that MYLAB will definitely improve science teaching and learning because it is easily accessible and easy to use by all learners and teachers.
- Thank you very much for dispelling the fear I had for doing practical work. My clumsiness suddenly disappeared. I would say that the next workshop will greatly improve my practical skills.

4 Discussion

4.1 Discussion of Formal and Informal Workshops

This section discusses the results of the workshops mentioned in the Results section. The results of the 24 pre-service teachers of a 2012 formal workshop (Box 3.2, example 1) show that the conceptual problems are more far-reaching than just not having the correct conceptual understanding. The student-teachers can only give general explanations e.g. the steps for teaching for conceptual change, without being able to give the specific chemical concept and chemical experiment they will use to teach in order to correct the misconception. They themselves have numerous misconceptions and lack sufficient theoretical chemistry knowledge to address the problems. Incorrect chemical reaction types (CRT), however, and incorrect driving forces can be addressed and were addressed by doing a series of experiments involving different reactions of the same type. Guided inquiry worksheets and short SSC experiments (MYLAB experiment 8 grade 10) lead the student-teachers to see the patterns and to identify the CRT and the driving force correctly. Nonetheless, students needed the hints and probing questions of the workshop presenter to nudge them in the right direction of finding an appropriate practical experiment to help learners comprehend the correct concept instead of the misconception.

The results of the 24 pre-service teachers of a 2011 formal workshop (Box 3.3, example 2) confirmed that these student-teachers had the same misconceptions about redox reactions as those reported by Horton (2007) and Taber (2002), namely: oxidation is the addition of oxygen and reduction is the removal of oxygen; if a reaction does not involve oxygen it is not oxidation; oxidation and reduction can occur independently and most often only oxidation is considered and reduction completely ignored. A series of experiments about redox reactions with guided-inquiry worksheets confronted the student-teachers with the fact that not all redox reactions contain oxygen. These chemistry misconceptions are not only found in South Africa but in many other contexts as well (Barke et al. 2009; Barker 2000; Mulford and Robinson 2002; Levy Nahum et al. 2004).

The results of example 3 (Box 3.4) for the same group of student-teachers indicated that "what they see is what they believe". The student-teachers get stuck on the macroscopic "explanation" of their observations, and this is the cause of most of the misconceptions that they have. The use of the chemistry triplet of easy transfer from macroscopic level to sub-microscopic level to symbolic level is not evident (Gabel 1999; Sirhan 2007). Lack of adequate subject knowledge is apparent (Kamau 2012; Marais and Mji 2009; Nbina 2012). The MYLAB kit's guided-inquiry worksheets are of great help to direct learners and students to the correct conclusions about a specific misconception. The worksheet contains questions guiding from direct observations to analysis, evaluation and critical thinking (Du Toit 2012a).

The results from the informal workshop for 48 in-service teachers held at the University of Venda were measured more on the affective level (Bybee et al. 2008): attitude, enjoyment, enthusiasm, co-operation and motivation. Teachers learned process skills by being assigned a specific task in the team. Teams with the most effective role assignments made the best progress. They had to take joint responsibility for the successful outcome of the assignment. With a more personal and relaxed approach, misconceptions could be changed to correct concepts; with extensive colleagues' cooperation, conceptual change could be brought about.

4.2 Discussion of the Materials

Three issues are considered in this section: the effectiveness of the MYLAB kit as a laboratory, its usefulness as intervention tool to address misconceptions, and its advantages to reach remote areas.

The MYLAB kits are ideal for resource constrained schools, as no physical resources are required from the school. No laboratories, expensive equipment, water or electricity is necessary, and not even classrooms (Tesfamariam et al. 2015). Everything is supplied in the kit. The only extra equipment is a 2-liter milk bottle with water, 2 empty plastic containers (one for water waste and one for paper waste) per kit, and one toilet roll (called "micro-towels") to dry apparatus and clean-up spills. Therefore, the availability, versatility and cost-effectiveness of SSC kits make them particularly successful in providing teaching assistance for under-resourced schools and as a tool to re-establish practical work. The only requirement is the willingness and enthusiasm of the teachers attending the workshops. The versatility of the kits makes it possible to address any concern or misconceptions the teachers have. The kits can be used easily and effectively to confront learners and teachers with the correct scientific phenomena through practical experiments. Moreover, learners usually observe experiments by watching the teacher perform them; they never have the opportunity to do the practical experiments themselves. The MYLAB kits address this gap by supplying not one but many 'laboratories', allowing learners to do the work themselves. This allows individual problems and misconceptions the learners might have to be identified, as learners learn better when doing the practical work themselves (Abdullah et al. 2009; Hofstein and Mamlok-Naaman 2007; Mafumiko 2008; Tobin 1990).

The value of the MYLAB kit as a tool to make chemistry available for the masses can be measured by its comparison to other small-scale endeavours. Most of the other researchers utilizing small scale apparatus only focus on a few experiments, i.e. gas exchange (El-Marsafy et al. 2011; Mattson et al. 2006), whilst others have very expensive components (Singh et al. 1999; Tantayanon 2005). The comprehensiveness of the MYLAB kits makes them very suitable for the range of experiments that schools need for all the different grades. Moreover, this also makes them excellent tools for addressing an assortment of misconceptions held by teachers and learners. No extra chemicals or apparatus are needed to teach for conceptual change when correcting misconceptions. All apparatus and all general chemicals are available to conduct experiments. Small apparatus and little amounts of chemicals makes experimental times short, therefore, a large number of experiments can be done in a limited time. To clarify misconceptions, numerous experiments can be done in a short time and a series of similar experiments to enable learners to discern a pattern can be conducted (Du Toit 2012a).

Because of the small size and comprehensiveness of the MYLAB kits, it was possible to take along approximately 50 kits when visiting numerous rural schools. Workshops in inaccessible places were possible. With highly portable kits, any school with or without resource barriers can be reached. Some of the areas where workshops were conducted were very remote, with only rutted, dirt, access roads. Moreover, these schools also have no shops or pharmacies within easy reachable distances to buy or replenish consumables.

4.3 Limitations of the Study

The following features need also to be taken into account for an overall evaluation of the conducted workshops:

- Workshops are presented on demand and no organized plan is followed.
- Some schools have little support from their district offices, and begged the author to come regularly once a quarter. Time, money and other obligations are the biggest deterrent to additional visits. Follow-up visits are almost impossible due to money and time constraints. The lack of follow-up visits can partly be solved with support groups in the local communities

- Lack of Department of Basic Education involvement (and thus government involvement) makes organized planning and support of at-risk schools impossible. This absence of involvement still happens despite a very positive report from the Quality Assurance Chief Directorate (Tholo et al. 2006).
- The presentation of workshops depends on the generosity of private institutions.
- The status quo of school problems is taken as part of the existing scenario, as workshop presenter cannot solve the social, physical and logistical problems of the school.
- All the learners and teachers are second language users.
- Only limited insight is gained in teachers' qualifications and PCK (pedagogical content knowledge) from pre-workshop questionnaires.
- The poor subject knowledge background of teachers and learners requires more time spent on support than it is possible during single workshops.

5 Conclusions

The following conclusions can be derived from what has been presented in the previous sections.

The MYLAB kits are quick and cost-effective replacements for traditional laboratories. The results from this study, and from other international studies (Tesfamariam et al. 2014 and 2015), bear witness to the effectiveness of the SSC kits. The post-workshop evaluations for all the approximately 260 workshops and all the feedback from workshop participants are positive. However, consideration must be given to the fact that the shift from having no opportunity for practical work to having individual opportunity for it is bound to produce positive results. Thus, the feedback also indicates the need and gratitude for the practical support provided.

This study demonstrated how SSC experiments can be used successfully to address and rectify misconceptions by teaching for conceptual change. The easiness with which experiments can be done, the help the worksheets provides, the comprehensiveness of the kit, and the short time it takes to complete an experiment, all contribute towards the ability of the MYLAB kit to address the misconceptions encountered in the class. As seen from the results, misconceptions are prevalent and an active way to address them is relevant. The MYLAB SSC kits are thus useful in teaching for conceptual change (Niaz 1995, 2006; Riordan 2012).

The value of small-scale chemistry to address misconceptions was proven and the possibility to re-establish practical work in diverse communities was demonstrated. Due to the compactness and low-cost of the kits it is possible to bring chemistry and science to the learner-population of a country.

Acknowledgements Thank you to Jean I. du Toit for critical reading and text editing of the article. Thank you to Marié J. du Toit for drawing the map and for critical reading of the article.

Appendices

Appendix A: List of the Names of the Places Indicated on the Map in Fig. 3.1, Where South African Workshops Were Held

Place	Province	Position
Bela-Bela	Limpopo	S24 53 04.3 E28 17 37.6
Bloemhof	North West	S27 39 06.5 E25 36 19.1
Boshof	Free State	S28 32 21.3 E25 14 10.2
Bothaville	Free State	S27 23 35.2 E26 36 58.3
Brits	North West	S25 38 06.1 E27 46 43.0
Bultfontein	Free State	S28 17 30.4 E26 09 11.9
Cape Town	Western Cape	S33 55 25.6 E18 25 24.0
Durban	Kwazulu-Natal	S29 51 25.2 E31 01 29.2
East London	Eastern Cape	S33 00 48.4 E27 54 15.1
Ganyesa	North West	S26 35 30.4 E24 10 15.1
Giyani	Limpopo	S23 18 34.4 E30 41 29.1
Johannesburg	Gauteng	S26 12 05.2 E28 02 43.7
Kimberley	Northern Cape	S28 44 16.1 E24 45 58.4
Kleinmond	Western Cape	S34 20 26.4 E19 01 51.8
Klerksdorp	North West	S26 52 13.8 E26 39 52.1
Kroonstad	Free State	S27 39 42.1 E27 14 04.5
Kuruman	Northern Cape	S27 27 43.7 E23 25 52.0
Ladybrand	Free State	S29 11 47.9 E27 27 24.7
Lichtenburg	North West	S26 09 04.7 E26 09 37.3
Mafikeng	North West	S25 51 52.3 E25 38 32.5
Middelburg	Mpumalanga	S25 45 57.9 E29 27 28.1
Mokopane	Limpopo	S24 11 01.4 E29 00 38.8
Monsterlus	Limpopo	S25 01 29.8 E29 43 52.2
Mookgophong	Limpopo	S24 31 09.2 E28 42 42.1
Moretele	North West	S27 19 21.0 E24 41 25.7
Phuthaditjhaba	Free State	S28 30 06.1 E28 49 07.0
Polokwane	Limpopo	S23 54 44.1 E29 27 12.8
Potchefstroom	North West	S26 42 56.9 E27 05 41.7
Pretoria	Gauteng	S25 44 45.6 E28 11 13.6
Relela	Limpopo	S23 40 50.9 E30 19 39.4
Rustenburg	North West	S25 40 06.6 E27 14 32.6
Saint Lucia	Kwazulu-Natal	S28 22 35.6 E32 24 46.9
Sasolburg	Free State	S26 48 58.0 E27 49 47.0
Secunda	Mpumalanga	S26 30 24.6 E29 11 59.5
Sekgopo	Limpopo	S23 37 06.3 E29 58 53.6
Sekhukhune	Limpopo	S24 45 11.3 E30 00 36.3

(continued)

Place	Province	Position
Stanford	Western Cape	S34 26 26.7 E19 27 30.9
Taung	North West	S27 31 40.0 E24 47 10.3
Thohoyandou	Limpopo	S22 57 07.3 E30 28 23.7
Tierkloof	North West	S27 03 20.4 E24 45 25.7
Turkey	Limpopo	S24 19 14.9 E30 31 43.7
Tzaneen	Limpopo	S23 49 36.7 E30 09 45.0
Virginia	Free State	S28 06 23.5 E26 52 03.8
Vredenburg	Western Cape	S32 54 25.9 E17 59 25.9
Vryburg	North West	S26 57 37.1 E24 43 35.7
Welkom	Free State	S27 58 50.4 E26 44 05.8
Winburg	Free State	S28 31 11.1 E27 00 41.4
Zeerust	North West	S25 32 40.4 E26 04 40.3

Appendix B: Results for Example 1 (Box 3.2) in the 2012 Workshop Described in the Text

Student	M1 of E1 Total marks (mark/12)	Q1 (mark/1)	Q2 (mark/1)	Q3 (mark/2)	Q4 (mark/8)	Number of misconceptions in answers
1	5(41.7%)	0	0	1	4	two
2	1(8.3%)	0	0	0	1	one
3	1(8.3%)	0	0	0	1	one
4	1(8.3%)	0	0	1	0	
5	1½(12.5%)	0	0	1	½	
6	7(58.3%)	1	1	0	5	one
7	2½(20.8%)	0	0	1	1½	
8	6(50%)	1	1	1	3	
9	0(0%)	0	0	0	0	two
10	4½(37.5%)	1	1	1	1½	one
11	½(4.2%)	0	0	½	0	two
12	2½(20.8%)	0	0	½	2	one
13	6(50%)	0	½	1	4½	
14	3(25%)	0	0	1	2	two
15	5(41.7%)	1	1	1	2	
16	2(16.7%)	0	1	1	0	two
17	3(25%)	0	0	1	2	
18	2(16.7%)	0	0	½	1½	
19	6(50%)	0	1	1	4	
20	4(33.3%)	1	1	0	2	

(continued)

Student	M1 of E1 Total marks (mark/12)	Q1 (mark/1)	Q2 (mark/1)	Q3 (mark/2)	Q4 (mark/8)	Number of misconceptions in answers
21	4½(37.5%)	0	0	1	3½	
22	5(41.7%)	0	0	1	4	one
23	9(75%)	1	1	1	6	
24	7(58.3%)	1	1	1	4	
Average	**3.7**	**0.29**	**0.40**	**0.73**	**2.29**	
	30.8%	**29%**	**40%**	**36.5%**	**28.6%**	

The symbols are defined in the caption of Box 3.2

Appendix C: Results from a 2011 Formal Workshops for Example 2 and Example 3[a]

Student	Total (mark/20)	E2 Q1 (mark/10)	E3 Q2 (mark/10)	E3 Q2 M1 (mark/2)	E3 Q2 M2 (mark/2)	E3 Q2 M3 (mark/2)	E3 Q2 M4 (mark/2)	E3 Q2 M5 (mark/2)
1	11.5	7.5	4	1	1	2	0	0
2	14.5	8	6.5	2	2	2	0.5	0
3	4.5	2.5	2	0	0	1	0	1
4	10.5	4.5	6	0	2	2	1	1
5	6	4.5	1.5	0	0.5	0	1	0
6	16.5	7	9.5	2	2	2	1.5	2
7	3	0	3	0	0	2	1	0
8	11	5.5	5.5	2	1	2	0	0.5
9	11.5	4	7.5	0.5	2	2	2	1
10	5	1	4	1	0	2	0	1
11	12	5	7	1	1	2	2	1
12	9	4.5	4.5	0	2	2	0	0.5
13	4.5	3	1.5	0	1	0	0	0.5
14	5.5	4	1.5	0	1	0	0	0.5
15	8.5	2	6.5	1	2	2	0.5	1
16	9.5	7	2.5	0	1	1	0	0.5
17	3.5	3	0.5	0	0	0.5	0	0
18	6	4.5	1.5	0	0.5	0	0.5	0.5
19	8	4	4	0	1	2	1	0
20	8.5	5	3.5	0.5	0	2	0.5	0.5
21	9.5	3	6.5	1	2	2	1	0.5
22	7.5	5	2.5	0	0	2	0.5	0
23	10	4.5	5.5	1	1	2	1	0.5
24	7.5	3.5	4	1	1	2	0	0
Average	8.48	4.27	4.21	0.58	1.00	1.52	0.58	0.52

(continued)

Student	Total (mark/20)	E2 Q1 (mark/10)	E3 Q2 (mark/10)	E3 Q2 M1 (mark/2)	E3 Q2 M2 (mark/2)	E3 Q2 M3 (mark/2)	E3 Q2 M4 (mark/2)	E3 Q2 M5 (mark/2)
Top marks				**3**	**7**	**17**	**2**	**1**
%	**42.4**	**42.7**	**42.1**	**29**	**50**	**76**	**29**	**26**

M1 is misconception 1, etc.
Q1 is question 1, etc.
[a]E1 is example 1, E2 is example 2, E3 is example 3

References

Abdullah, M., Mohamed, N., & Ismail, Z. H. (2009). The effect of an individualized laboratory approach through microscale chemistry experimentation on students' understanding of chemistry concepts, motivation and attitudes. *Chemistry Education Research and Practice, 10*(1), 53–61.

Adams, R. (2014). *Practical work must remain part of science A-levels, say experts*. Retrieved from https://www.theguardian.com/education/2014/jan/17/science-practical-work-experts-lambast

Ali, T. (2012). A case study of the common difficulties experienced by high school students in chemistry classroom in Gilgit-Baltistan (Pakistan). *SAGE Open*, 2158244012447299.

Arends, F., Winnaar, L., Zuze, T., Visser, M., Reddy, V., Prinsloo, C., . . . Rogers, S. (2015). Beyond benchmarks: What twenty years of TIMSS data tell us about South African education. Cape Town: HSRC Press.

Barke, H., Hazari, A., & Yitbarek, S. (2009). *Misconceptions in chemistry: Addressing perceptions in chemistry*. Berlin: Springer.

Barker, V. (2000). *Beyond appearances: students' misconceptions about basic chemical ideas*. A report prepared for the Royal Society of Chemistry. Retrieved from http://modeling.asu.edu/modeling/KindVanessaBarkerchem.pdf.

Beaton, A. E. (1996). *Mathematics Achievement in the Middle School Years. IEA's Third International Mathematics and Science Study (TIMSS)*. Chestnut Hill: ERIC.

Bybee, R. W., Carlson-Powell, J., & Trowbridge, L. W. (2008). *Teaching secondary school science: Strategies for developing scientific literacy*. Harlow: Pearson/Merrill/Prentice Hall Columbus.

Du Toit, M. H. (2006). *Preparing teachers for reform – Chemistry teacher education: Teaching Chemistry through workshops for teachers – a report on a successful program*. Unpublished report: North-West University.

Du Toit, M. H. (2012a). *MYLAB worksheets grade 8–12: New CAPS*. Unpublished report: North-West University.

Du Toit, M. H. (2012b). SEDIBA – A Teachers' Training Success Story (1996–2012 and beyond). *EC2E2N NewsLetter, 13*.

Du Toit, M. H. (2014). *MYLAB workshops 2014*. Unpublished report: North-West University.

Du Toit, M. H. (2015). *MYLAB workshops 2015*. Unpublished report: North-West University.

Du Toit, M. H. (2016). *MYLAB workshops in Tzaneen region 2016*. Unpublished report: North-West University.

El-Marsafy, M., Schwarz, P., & Najdoski, M. (2011). *Microscale chemistry experiments using water and disposable materials*. Nuzha: Kuwait Chemical Society.

Gabel, D. (1999). Improving teaching and learning through chemistry education research: A look to the future. *Journal of Chemical Education, 76*(4), 548.

Heeralal, P. (2014). Barriers experienced by natural science teachers in doing practical work in primary schools in Gauteng. *International Journal of Educational Sciences, 7*(3), 795–800.

Hewson, P. W. (1992). *Conceptual change in science teaching and teacher education* (Paper presented at the a meeting on "Research and Curriculum Development in Science Teaching," under the auspices of the National Center for Educational Research, Documentation, and Assessment, Ministry for Education and Science, Madrid, Spain).

Hofstein, A., & Mamlok-Naaman, R. (2007). The laboratory in science education: the state of the art. *Chemistry Education Research and Practice, 8*(2), 105–107.

Horton, C. (2007). Student alternative conceptions in chemistry. *California Journal of Science Education, 7*(2), 1–78.

Kamau, D. M. (2012). *A study of the factors responsible for poor performance in Chemistry among secondary school students in Maragua District.*

Kibirige, I., Rebecca, M. M., & Mavhunga, F. (2014). Effect of Practical Work on Grade 10 Learners' Performance in Science in Mankweng Circuit, South Africa. *Mediterranean Journal of Social Sciences, 5*(23), 1568.

Kyle, W., & Shymansky, J. A. (1989). Enhancing learning through conceptual change teaching. *NARST News, 31*(3), 7–8.

Levy Nahum, T., Hofstein, A., Mamlok-Naaman, R., & Bar-Dov, Z. (2004). Can final examinations amplify students' misconceptions in chemistry? *Chemistry Education Research and Practice, 5*(3), 301–325.

Mafumiko, F. M. (2008). The potential of micro-scale chemistry experimentation in enhancing teaching and learning of secondary chemistry: Experiences from Tanzania Classrooms. *NUE Journal of International Cooperation, 3*, 63–79.

Marais, F., & Mji, A. (2009). *Factors contributing to poor performance of first year chemistry students*. INTECH Open Access Publisher.

Martin, M. O., Mullis, I. V., & Chrostowski, S. J. (2004). *TIMSS 2003 technical report: Findings from IEA's trends in international mathematics and science study at the fourth and eighth grades*. Chestnut Hill: ERIC.

Martin, M. O., Mullis, I. V., Foy, P., & Stanco, G. M. (2012). *TIMSS 2011 international results in science*. Boston: ERIC.

Mattson, B., Anderson, M. P., & Mattson, S. (2006). *Microscale gas chemistry*. Norwalk: Educational Innovations.

Millar, R., & Abrahams, I. (2009). Practical work: making it more effective. *School Science Review, 91*(334), 59–64.

Mulford, D. R., & Robinson, W. R. (2002). An inventory for alternate conceptions among first-semester general chemistry students. *Journal of Chemical Education, 79*(6), 739.

Nbina, J. B. (2012). Analysis of poor performance of senior secondary students in chemistry in Nigeria. *African Research Review, 6*(4), 324–334.

Niaz, M. (1995). Cognitive conflict as a teaching strategy in solving chemistry problems: a dialectic–constructivist perspective. *Journal of Research in Science Teaching, 32*(9), 959–970.

Niaz, M. (2006). Facilitating chemistry teachers' understanding of alternative interpretations of conceptual change. *Interchange, 37*(1–2), 129–150.

Potgieter, M., & Davidowitz, B. (2011). Preparedness for tertiary chemistry: multiple applications of the Chemistry Competence Test for diagnostic and prediction purposes. *Chemistry Education Research and Practice, 12*(2), 193–204.

Riordan, J.-P. (2012). Strategies for Conceptual Change in School Science. *Educational Psychology, 92*(3), 413–425.

Singh, M. M., Szafran, Z., & Pike, R. M. (1999). Microscale chemistry and green chemistry: Complementary pedagogies. *Journal of Chemical Education, 76*(12), 1684.

Sirhan, G. (2007). Learning difficulties in chemistry: An overview. *Journal of Turkish Science Education, 4*(2), 2.

Spaull, N. (2013). *South Africa's education crisis: The quality of education in South Africa 1994–2011*. Report Commissioned by CDE, (pp. 1–65).

Taber, K. (2002). *Chemical misconceptions: Prevention, diagnosis and cure* (Vol. 1). London: Royal Society of Chemistry.

Tantayanon, S. (2005). *Small scale laboratory: Organic Chemistry at University Level*. Bangkok: Chulalongkorn University. Retrieved from http://www.unesco.org/science/doc/Organi_chem_220709_FINAL.pdf.

Tesfamariam, G., Lykknes, A., & Kvittingen, L. (2014). Small-scale chemistry for a hands-on approach to chemistry practical work in secondary schools: Experiences from Ethiopia. *African Journal of Chemical Education, 4*(3), 48–94.

Tesfamariam, G. M., Lykknes, A., & Kvittingen, L. (2015). 'Named small but doing great': An investigation of small-scale chemistry experimentation for effective undergraduate practical work. *International Journal of Science and Mathematics Education*, 1–18.

Tholo, J. A. T., Bathobame, S. T., Swaratlhe, N. W., Dolo, J., Molema, M., Ncoane, J., ... Masilela, B. (2006). *North West Province: Quality assurance chief directorate report evaluation of the Chemistry student lab project*. Unpublished report.

Tobin, K. (1990). Research on science laboratory activities: In pursuit of better questions and answers to improve learning. *School Science and Mathematics, 90*(5), 403–418.

Chapter 4
The Systemic Approach to Teaching and Learning Organic Chemistry (SATLOC): Systemic Strategy for Building Organic Chemistry Units

Ameen F. M. Fahmy

1 Introduction

Systemic Approach to Teaching and Learning (SATL) is a new way of teaching and learning, based on the idea that nowadays everything is related to everything globally. Students should not learn isolated facts by rote, but they should be able to connect concepts and facts in an internally logical context. Taagepera and Noori (2000) tracked the development of student's conceptual understanding of organic chemistry during a one-year sophomore course. They found that the student's knowledge base increased as expected, but their cognitive organization of the knowledge was surprisingly weak. The authors concluded that instructors should spend more time making effective connections, helping students construct a knowledge space based on general principles. Fahmy and Lagowski (1999, 2003) Lagowski and Fahmy (2011), Fahmy (2017), have designed, implemented, and evaluated the systemic approach to teaching and learning chemistry (SATLC) in Egypt, since (1998). SATLC is a method of arranging concepts in such a way that the relationships between various concepts and issues are made clear. SATL methods have been shown, empirically, to be successful in helping students learn in a variety of settings: pre-college, college, and graduate. They facilitate deep learning of the subject by the students and enable them to grasp and correlate concepts and ideas that have been gathered from a variety of different disciplines (Fahmy and Lagowski 2011), Nazir and Naqvi (2012). A number of statistical studies involving student achievement indicate that students taught with SATL methods by teachers trained in those methods achieve significantly higher levels than those taught by standard linear methods of instruction (Fahmy and Lagowski 2003). A one-semester course (16 lectures, 32 hours) on aromatic chemistry using the SATL technique was organized

A. F. M. Fahmy (✉)
Department of Chemistry, Faculty of Science, Ain Shams University, Cairo, Egypt
e-mail: afmfahmy42@yahoo.com

L. Mammino, J. Apotheker (eds.), *Research in Chemistry Education*,
https://doi.org/10.1007/978-3-030-59882-2_4

and taught successfully to 28 2nd year students at Menoufia University during the academic year 2000/2001 (Fahmy et al. 2001).

2 SATL Strategy for Building Units in Organic Chemistry

2.1 *General Outline*

Within the work on SATL building strategy, the linearly based teaching units in chemistry are converted to systemically-based units according to the following general building strategy (Fahmy and Lagowski 2011, 2013), Nazir and Naqvi (2012):

Step 1: The systemic aims and the operational objectives for the unit should be defined in the frame of national standards.

Step 2: The knowledge from previous units (concepts, facts, reaction types and skills) needed as prerequisites for teaching a given unit should be tabulated in a list.

Step 3: The content of the linearly-based unit should be analysed into concepts, facts, and reaction types, and mental and experimental skills should be identified.

Step 4: A diagram illustrating linear relations among the concepts of the unit should be drawn.

Step 5: The linear diagram is modified by putting the (√) sign on the already known relationships between concepts, and the (X) sign on remaining unknown linear relationships.

Step 6: The final linear diagram should be modified to a systemic diagram SD0 by adding unknown relations between the concepts. SD0 is known as the starting point of teaching the unit.

Step 7: The students are expected to closely follow the content of the taught unit, step by step.

The unit starts with SD0 and ends with a terminal systemic (SDf) in which all the relationships between concepts are identified. In going from SD0 to SDf, the student will crossover several systemics with known and unknown relationships, like SD1, SD2, …..etc. (Fahmy and Lagowski 2011, 2013).

2.2 *Scenario of Building a Unit on Aromatic Chemistry: (Benzene and Related Compounds)*

With reference to the just outlined steps of the SATL general strategy for building organic chemistry units, this section presents the specific scenario for building a unit on systemic aromatic chemistry (Fahmy and Lagowski 2011).

Steps 1–3: These steps follow the general building strategy outlined for them in the previous subsection.

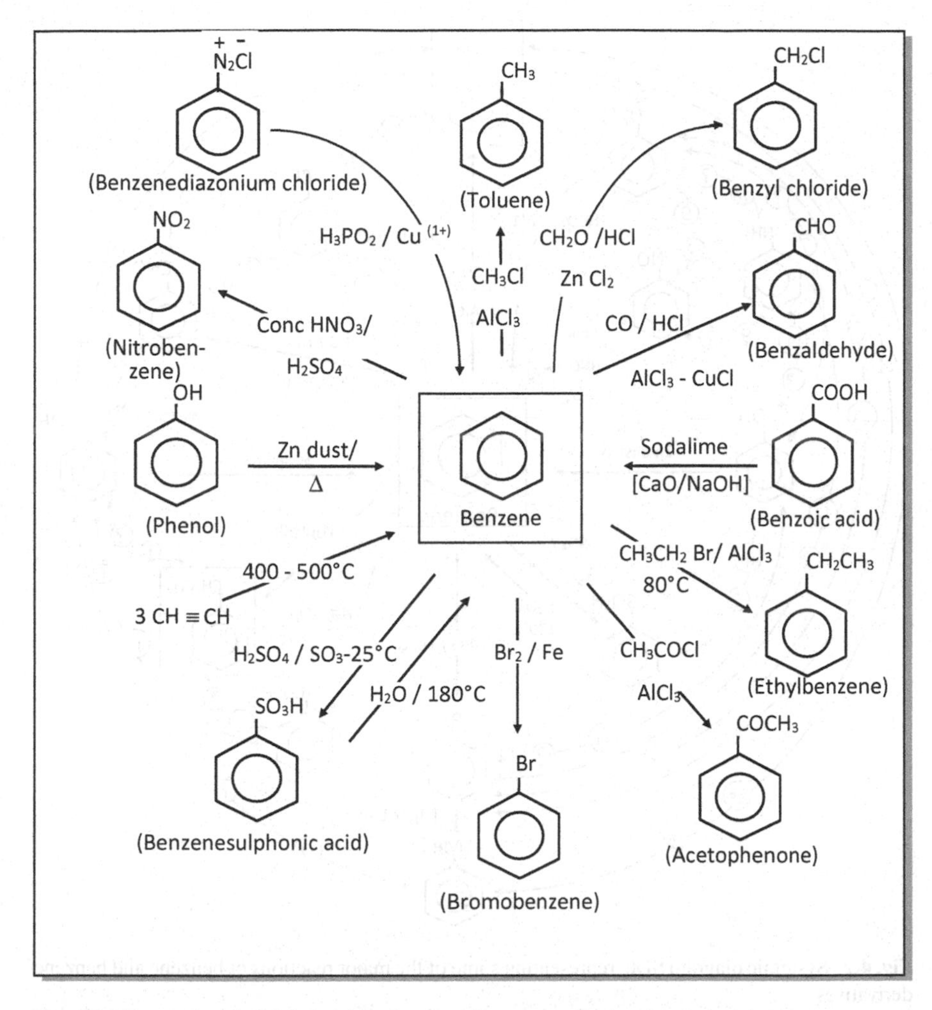

Fig. 4.1 Linear chemical relationships between benzene and related compounds

Step 4: After the students study the synthesis and reactions of benzene, they can draw the diagram shown in Fig. 4.1 to summarize this information. The diagram shows that all the involved chemical relations are linear and separated relations.

Step 5: The linear diagram in Fig. 4.1 can be transformed by students into a systemic diagram SD0, shown in Fig. 4.2. This diagram shows that the individual relationships of the compounds suggested to be synthesized from benzene in Fig. 4.1 (alkyl benzenes, nitrobenzene, halo-benzenes, phenols, aromatic alcohols, benzaldehyde, acetophenone, …….), can be systemically interconnected by adding the unknown chemical relations between them. SD0 is known as the starting point for teaching the unit (Fahmy and Lagowski 2011). It shows the unknown chemical relations (1–20) between aromatic compounds, considering the pairs of compounds listed in Table 4.1. These relations will be clarified later during the study of the unit.

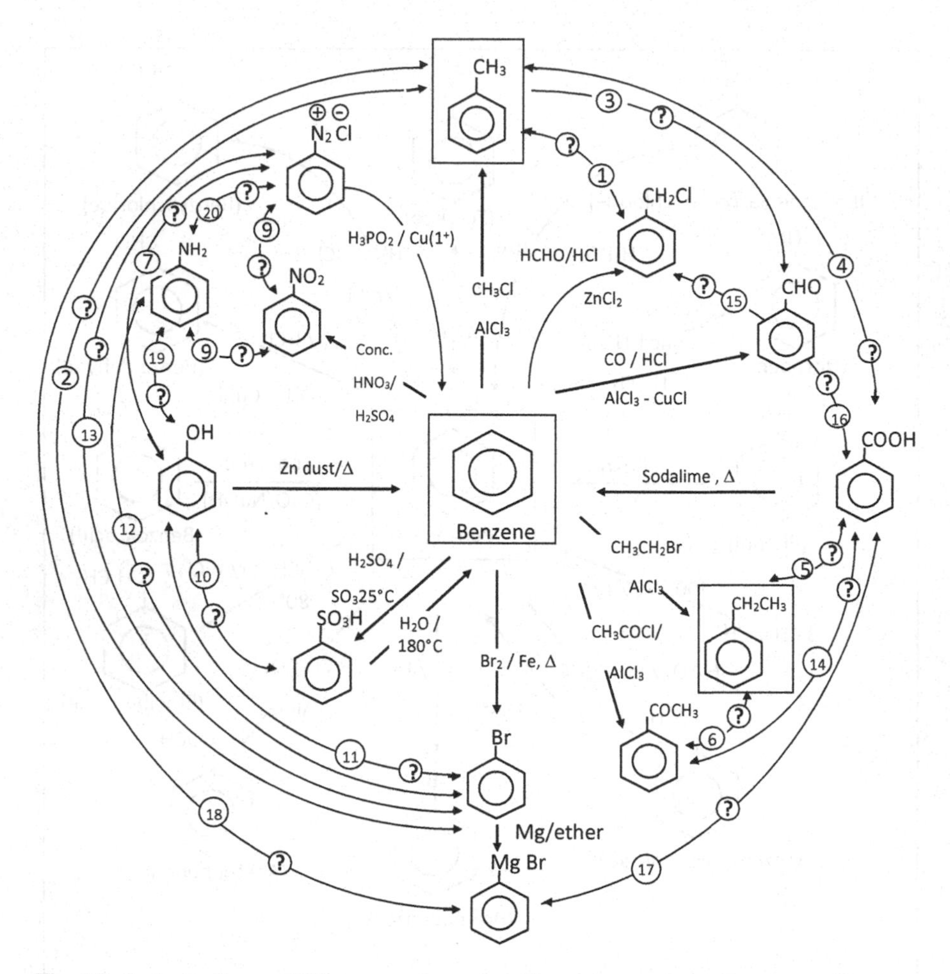

Fig. 4.2 Systemic diagram SD0, representing some of the major reactions of benzene and benzene derivatives

Step 6: **Building SD1 from SD0.** After all possible synthetic routes and reactions of alkyl benzenes are discussed in the classroom, and after the recognition of the relevant chemical relations, students can improve on the systemic diagram SD0 and build another systemic diagram, (SD1, Fig. 4.3) by adding chemical relations 1–6 and 18.

Step7: Building SD2 from SD1. After sufficient progress in teaching the unit via guided classroom discussions about the synthesis and reactions of halogen derivatives of aromatic hydrocarbons, the systemic diagram SD1 can be improved by students into systemic diagram SD2 (Fig. 4.4) by adding the chemical relations defined in Table 4.2. Chemical relations 7–10, 14, 16, 17 and 19 are still unknown and will be defined during our study of the rest of the aromatic chemistry unit.

Table 4.1 Unknown chemical relations between aromatic compounds, to be shown on building systemic diagram SD0

No.	Compounds to be related	Chemical relations
1	Toluene and benzyl chloride	?
2	Toluene and bromobenzene	?
3	Toluene and benzaldehyde	?
4	Toluene and benzoic acid	?
5	Ethylbenzene and benzoic acid	?
6	Acetophenone and ethyl benzene	?
7	Phenol and benzenediazonium chloride.	?
8	Nitrobenzene and benzene-diazonium chloride.	?
9	Nitrobenzene and aniline.	?
10	Phenol and benzenesulphonic acid.	?
11	Bromobenzene and phenol.	?
12	Bromobenzene and aniline	?
13	Bromobenzene and benzene-diazonium chloride	?
14	Acetophenone and benzoic acid	?
15	Benzyl chloride and benzaldehyde	?
16	Benzaldehyde and benzoic acid	?
17	Phenyl magnesium bromide and benzoic acid	?
18	Phenyl magnesium bromide and toluene	?
19	Phenol and aniline	?
20	Aniline and benzenediazonium chloride	?

Step.8: Building SD3 from SD2. After studying the synthesis and reactions of benzenesulphonic acids, the student can improve SD2 to SD3 (Fig. 4.5) by adding the reactions of sulphonic acids with NaOH (relation 10), and KSH, KCN, $NaNH_2$. Chemical relations 7–9, 14, 16, 17 and 19) are still unknown.

Step 9: Building SD4 from SD3. After studying the synthesis and chemical reactions of aromatic nitro-compounds, diazonium salts and aniline, the student can improve the systemic diagram SD3 to SD4 (Fig. 4.6) by adding the chemical relations listed in Table 4.3. Chemical relations 14, 16 and 17 are still unknown.

Step10: Building SD5 from SD4. After studying of synthesis and chemical reactions of aromatic aldehydes, ketones and acids, the student can modify the systemic diagram SD4 to build SDF (Fig. 4.7) by adding the chemical relations listed in Table 4.4.

In the SDF diagram, all chemical relations between benzene and its related compounds are clarified, and the diagram constitutes the end point of teaching the unit. At this point, new ideas and connections begin to arise in students' minds in a systemic pattern. This will opens the door for new approaches towards synthetic organic chemistry. In addition, it is possible to assess the students' achievements in aromatic chemistry after each stage of the learning unit via Systemic Assessment

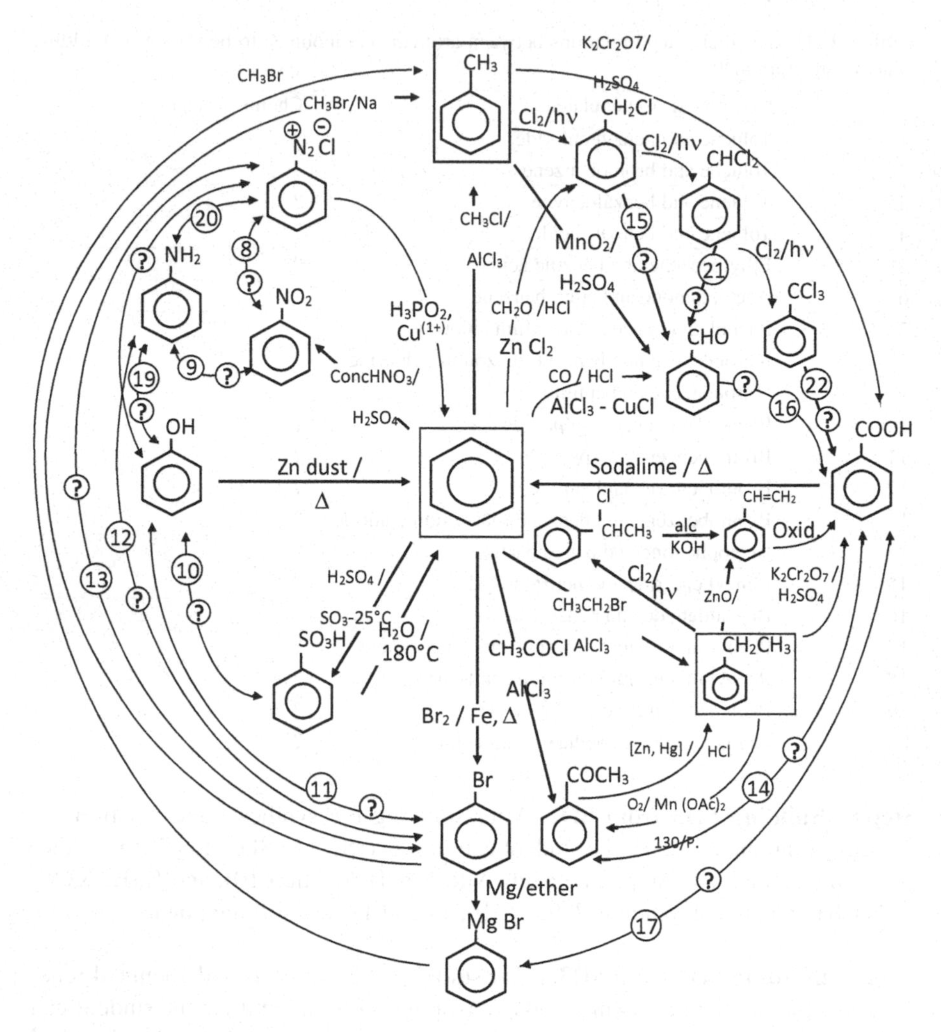

Fig. 4.3 Systemic diagram SD1

Questions [**SAQ, s**] on the systemic diagrams from **SD0** to **SD5** (Fahmy and Lagowski 2012).

3 Conclusions

The following conclusions were reached after the experimentation of SATLOC (Fahmy and Lagowski 2011, 2012):

- SATLOC improves the students' ability to view organic chemistry from a more global perspective.

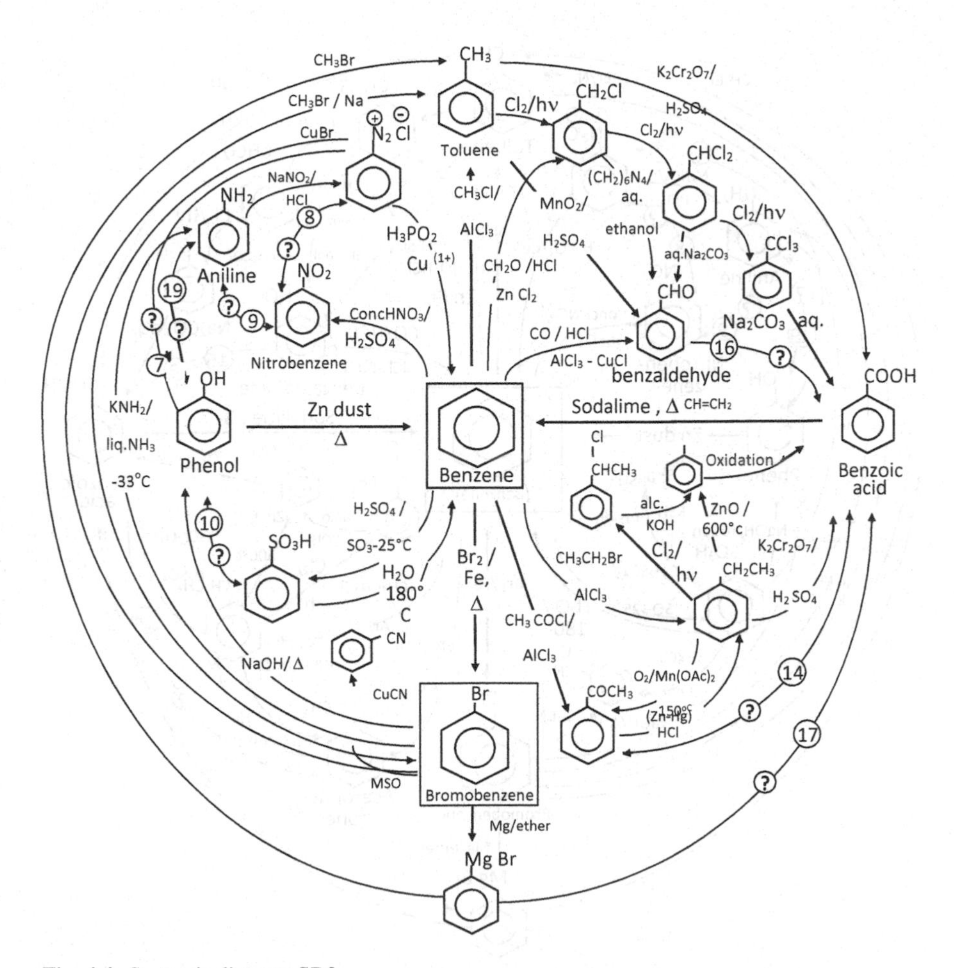

Fig. 4.4 Systemic diagram SD2

Table 4.2 Chemical relations considered to build systemic diagram SD2 from systemic diagram SD1

No.	Chemical relations	Known?
11	Bromobenzene to phenol [NaOH, heat]	✓
12	Bromobenzene to aniline [KNH_2/liq. NH_3]	✓
13	Benzenediazonium chloride to bromobenzene [CuBr]	✓
15	Benzyl chloride to benzaldehyde	✓
20	[$(CH_2)_6N_4$/aq.alc.]	✓
21	Aniline to benzenediazonium chloride	✓
22	($NaNO_2$/HCl) Benzal chloride to benzaldehyde [aq. Na_2CO_3] Benzotrichloride to benzoic acid [i) aq.Na_2CO_3, ii) HCl]	

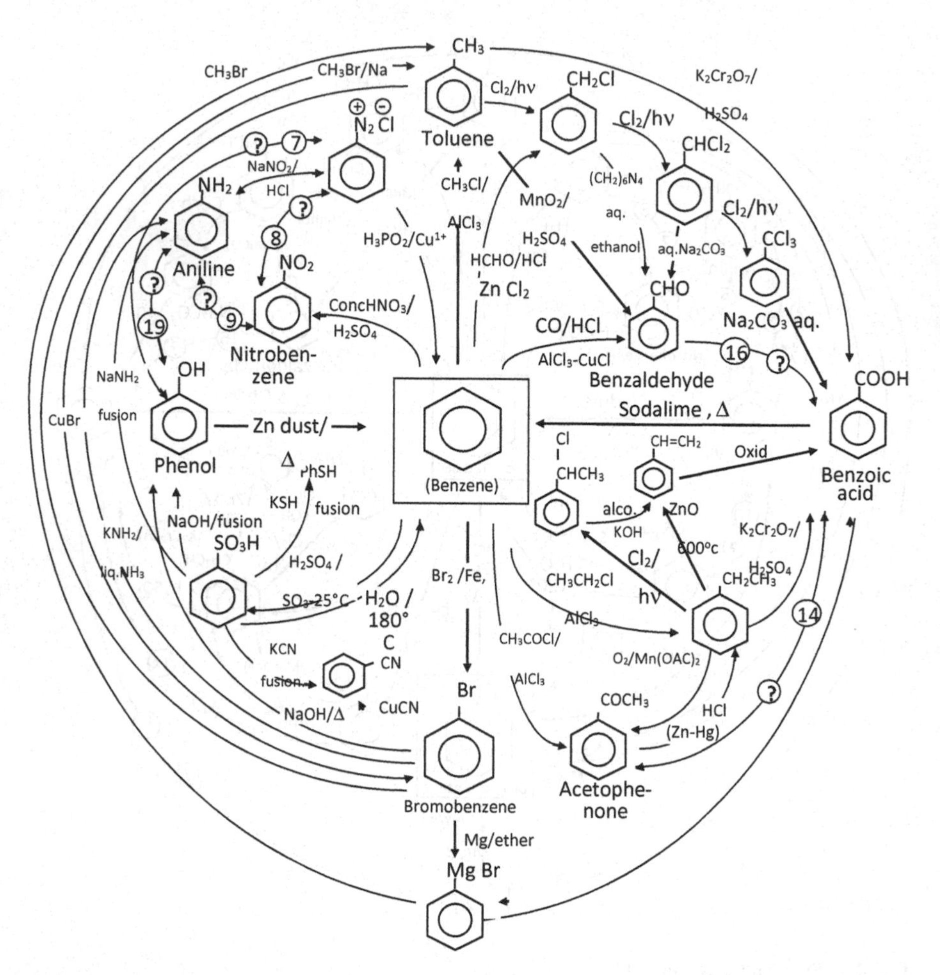

Fig. 4.5 Systemic diagram SD3

- SATLOC increases students' ability to learn the subject matter in a greater context.
- SATLOC helps the students develop their own mental framework at higher-level of cognitive processes (application, analysis, and synthesis).

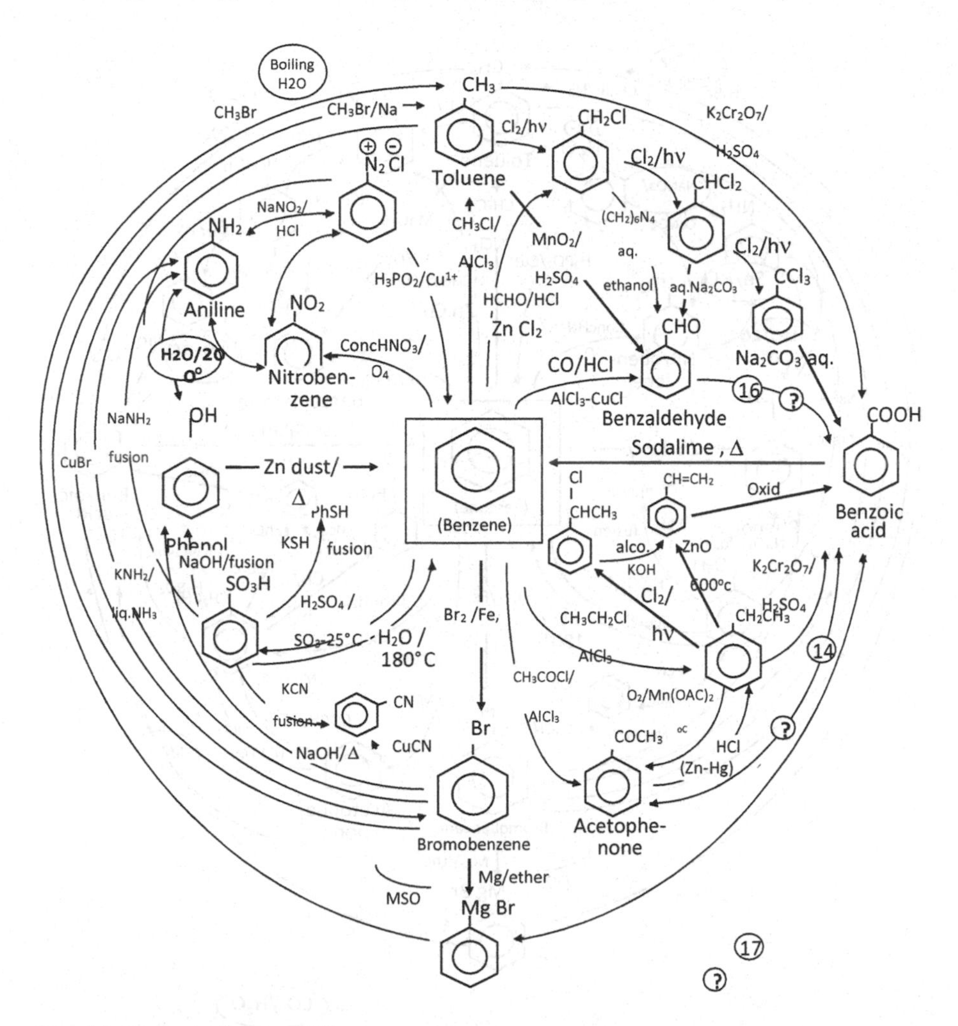

Fig. 4.6 Systemic diagram SD4

Table 4.3 Chemical relations considered to build systemic diagram SD4 from systemic diagram SD3

No.	Chemical relations	Known?
8	Benzenediazonium chloride to nitrobenzene ($NaNO_2$/Cu NO_2)	✓
7	Benzenediazonium chloride to phenol (boiling water)	✓
9	Nitrobenzene to aniline (reduction Sn / HCl)	✓
19	Aniline to phenol (H_2O, 200°).	✓

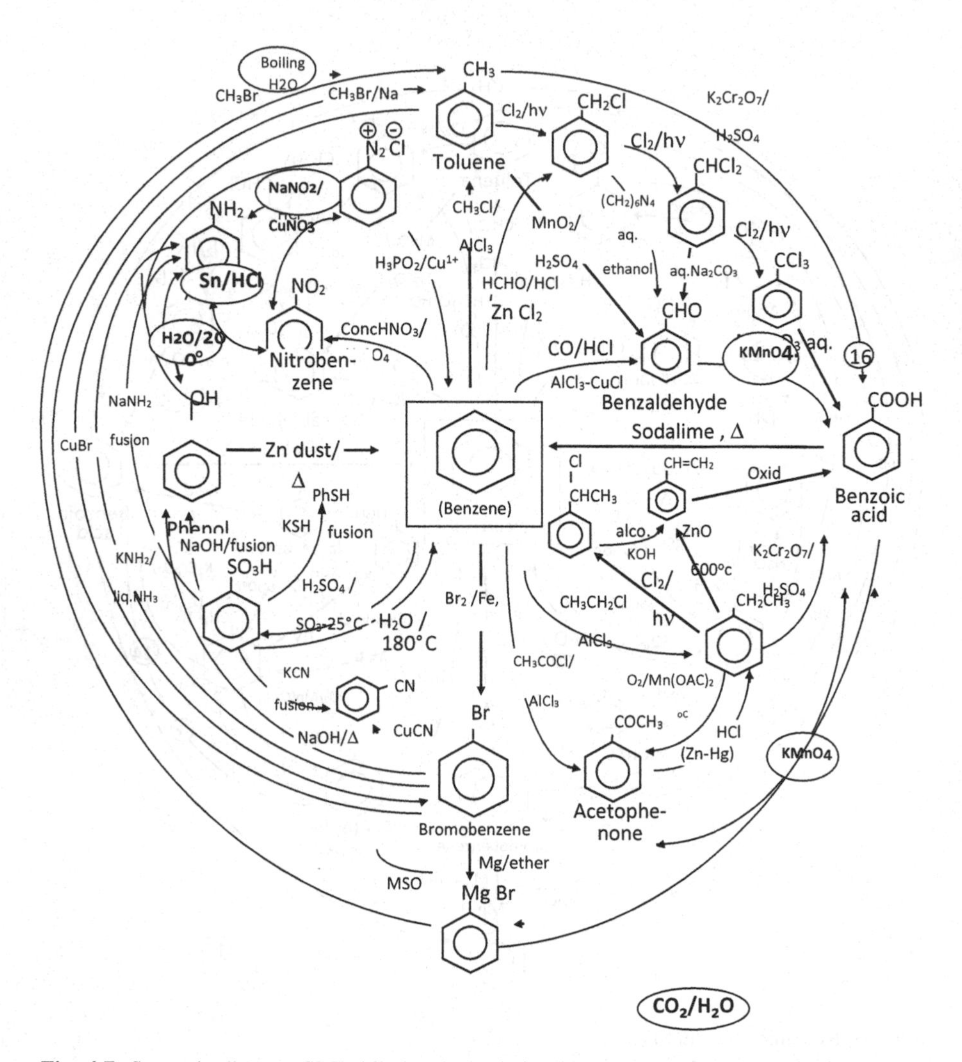

Fig. 4.7 Systemic diagram SDF. All chemical relationships between benzene and related compounds are clarified

Table 4.4 Chemical relations considered to build systemic diagram SDF from systemic diagram SD4

No	Chemical relation	Known?
14	Acetophenone to benzoic acid ($KMnO_4$)	√
16	Benzaldehyde to benzoic acid(KMnO4)	√
17	Phenyl magnesium bromide to benzoic acid (CO_2, H_2O)	√

References

Fahmy, A. F. M. (2017). The systemic approach to teaching and learning chemistry [SATLC]: A 20 years review. *AJCE, 7*(3), Special Issue 2–44.

Fahmy, A. F. M., & Lagowski, J. J. (1999). The uses of systemic approach to teaching and learning in 21st century. *Pure and Applied Chemistry, 71*(5), 859–863. [15th ICCE, Cairo, Egypt, August, 1998].

Fahmy, A. F. M., & Lagowski, J. J. (2003). Systemic approach to teaching and learning chemistry [SATLC] in Egypt [1998–2011]. *AJCE, 2*(2), 92–97.

Fahmy, A. F. M., & Lagowski, J. J. (2011). The systemic approach to teaching and learning (SATL): Operational steps for building teaching units. *AJCE, 1*(1), 62–80.

Fahmy, A. F. M., & Lagowski, J. J. (2012). Systemic assessment as a new tool to assess the student learning in chemistry using SATL methods: Systemic true false[STF], and Systemic sequencing [SSQs], Questions, types. *AJCE, 2*(2), 66–78.

Fahmy, A. F. M., & Lagowski, J. J. (2013). The systemic approach to teaching and learning heterocyclic chemistry [SATLHC]: Operational steps for building teaching unites in heterocyclic chemistry. *AJCE, 3*(2), 39–56.

Fahmy, A. F. M., Hashem, A. I., & Kandil, N. (2001). *Systemic approach in teaching and learning aromatic chemistry*. Cairo: Science Education Center.

Lagowski, J. J., & Fahmy, A. F. M. (2011). The systemic approach to teaching and learning, [SATL]: A 10-year's review. *AJCE, 1*(1), 29–47.

Nazir, M., & Naqvi, I. (2012). Designing of lectures through Systemic Approach to Teaching and Learning, a Model for (SATL) methodology. *Pakistan Journal of Chemistry, 2*(1), 46–57.

Taagepera, M., & Noori, S. (2000). Mapping student thinking patterns in learning. *Journal of Chemical Education, 77*, 1224.

Chapter 5
Are Our Students Learning and Understanding Chemistry as Intended? Investigating the Level of Prior Knowledge of UNIVEN Students for the Second Year Inorganic Chemistry Module

Malebogo A. Legodi

1 Introduction

Chemistry is an example of what Neumann (2001) describes as a “big science”; it is a hard pure discipline. In the teaching of hard pure disciplines, the ‘oldest’ knowledge (the knowledge established since long) is presented at the lower levels of instruction, as it constitutes the basic knowledge, whereas more current or advanced knowledge is taught at more advanced levels, such as senior undergraduate level (Neumann 2001).

Learning chemistry, like any other science, is a cumulative process where the teaching and learning of new information/concepts builds on what students already know (Neumann 2001; Ozmen 2004; Emondson 2005; Zoller 1990). What the students learn depends on their interpretation, because they bring to lessons pre-existing (alternative) conceptions about scientific phenomena (Barke et al. 2009; Palmer 1999, 2001; Taber 2000 in Ozmen 2004). These alternative conceptions can influence the way in which meaning is constructed in students’ minds and at times interfere with the learning of correct scientific principles or concepts (Ozmen 2004). The term ‘misconceptions’ is used synonymously with pre-existing or alternate conceptions throughout this document, inasmuch as they are different from the established concepts as accepted in a given field.

Misconceptions can be defined as ideas students have about concepts, which are inconsistent with scientific conceptions (Ozmen 2004). They reflect the complex nature of the multiple causes of students’ erroneous conceptions. There is a difference between lack of knowledge or concept and misconception. A lack of concept or knowledge can be remedied with instructions and subsequent learning, while

M. A. Legodi (✉)
Department of Chemistry, University of Venda, Thohoyandou, South Africa
e-mail: Malebogo.Legodi@univen.ac.za

L. Mammino, J. Apotheker (eds.), *Research in Chemistry Education*,
https://doi.org/10.1007/978-3-030-59882-2_5

misconceptions need unlearning before new material can be learned. As such, methods for eliminating misconceptions and for remedying lack of concept may differ considerably (Hasan et al. 1999).

If students fail to acquire the appropriate knowledge, they may have difficulties participating in the disciplinary discourses (Biggs 1999). This will make it difficult for students to acquire problem-solving skills, ways of thinking and working in the discipline (Biggs 1999). Furthermore, after graduation, they may not be able to deal with the ever changing and complex challenges in the global and information-based society (Biggs 1999).

The process of students making their own meaning is the basis of the constructivist approach prevalent in science learning. This approach assumes that the knowledge and understanding of concepts arise from the variety of contacts with the physical and social world, or as a result of personal experience, interaction with teachers, other people or through the media (Gilbert and Zylberstajn 1985 in Ozmen 2004; Palmer 2001). Since misconceptions can interfere with quality chemistry learning, it is incumbent on teachers to determine ways of prevention or remediation (Schoon and Boone 1998). The teacher should strive to help students achieve good, chemically accurate understanding of the concepts.

The misconceptions can be deeply rooted in students' minds because they have been acquired over years and are, in their nature, difficult to change. They can, thus, form part of the students' core beliefs (Ozmen 2004). Factors such as teaching approach, assessment criteria, use of models, internet-based teaching (ICTs) and so forth, can also contribute to the creation of misconceptions in students' minds (Kind 2004; Nahum et al. 2004; Boo 1998).

The majority of students at the University of Venda are African and from rural areas (where educational institutions are under-resourced) and English is not their mother-tongue. A significant percentage of students, therefore, show poor English language proficiency and thus experience difficulties understanding lectures, let alone the textbooks and other chemistry literature, or the disciplinary terminology. These difficulties could lead to misconceptions. Furthermore, for these students, articulation gap is expected, which further increases the barriers to learning. Articulation gap is defined as the mismatch between the learning requirements of the current module and the skills and competencies acquired from the previous prerequisite module (Fischer and Scott 2011). It means that there is a difference between the skills and competencies that are required for the module and those possessed by the students. This may be due to the mismatch between the minimum requirements for the module and the academic preparedness needed for success in the course (ibid). The presence of articulation gap causes a discontinuity in the students' learning process as they move from a previous level to a more advanced one.

In an effort to unearth misconceptions held by students as they begin the 1st semester Inorganic Chemistry II course (course code: CHE 2521), this study analyses responses to a diagnostic test based on high school and 1st year chemistry syllabi. The approach was aimed at allowing the students to express their views on different chemistry subjects (National Research Council 1997). Since a qualitative research approach was chosen, questions and discussions are used to probe misconceptions (National Research Council 1997). Some misconceptions have been discovered by asking students to sketch or describe some object or scientific phenomenon (National Research Council 1997). The use of a diagnostic test based on the preceding modules and given at the beginning of the course can be another way of inducting students into the chemistry discipline. It gives the new students an idea of what sort of preparatory work or level of skills is required for enrolment into the course. The test will also give the teacher an idea of how well prepared the students are for the course. The literature reports that in all sciences, including chemistry, phenomena are associated with some form of students' misconceptions. The literature reports studies on students' misconceptions related to chemistry concepts, and investigates their possible causes also through blended learning and examinations (Ozmen 2004; Tan et al. 2002). The current study reports on possible misconceptions implied by the responses to the diagnostic test based on hybridization, electronegativity, octet rule, bonding, periodic table, electronic configuration, Lewis structures and resonance. These concepts were selected as those for which incorrect answers were encountered frequently and which, therefore, posed challenges for the majority of students enrolled for undergraduate chemistry modules.

2 Motivations for the Study

The following observations from interaction with students are some of the factors that have motivated the study:

- General fear/negative perceptions from the students, e.g. "Chemistry is difficult"
- Widely spread memorization, which leads to cheating in tests and examinations
- Poor conceptual understanding of scientific principles/concepts
- Lack of interest in pursuing post-graduate studies (concerns the majority of students)
- High failure rate (low class average marks)
- Lack of participation in class
- Answering questions that were not asked
- Frequent cheating in tests and examinations (by illegal material or copying from each other)
- Difficulties in constructing coherent scientific arguments
- Reproduction of materials from the notes, textbooks and model answers in tests and examinations.
- Desperate need for model answers, that students memorize
- Few students own textbooks

The above-mentioned points were perceived as out of the ordinary by the author and prompted the current study as a way of finding out the possible underlying students' learning issues.

3 Methodology

The subjects of the research were 80 students of the University of Venda, majoring in chemistry, in their second year during the 2015 academic year, and who had enrolled for the CHE 2521 module. A qualitative research was carried out to identify their level of preparedness and possible misconceptions. The instrument used was a set of nine discussion questions as part of diagnostic test. The test contained no multiple choice questions or questions requiring one-word answers, to minimize the chances of guessing. Furthermore, the instructions informed that the test would not count for formal assessment. The author believes that, under these circumstances, the students would gather that only their honest opinions and impressions were sought, without consequences for them. Due to the created atmosphere of lack of perceived personal incentives or direct consequences, the author was confident that the students would be free to give their honest opinions. The students with lack of knowledge on particular questions were expected to leave blank spaces or to indicate that they did not know the answer. The majority of students answered most of the questions and no totally blank answer sheet was returned. The absence of response to certain questions was interpreted as due to lack of knowledge of the given concept, since there would be no perceived reason or incentive to give misleading answers.

The responses recorded in the present work were selected from the answers to questions in the diagnostic test (see Appendix A). The answers were analysed to identify module-wide misconceptions held by students of that group. The misconceptions are identified by analysing the meaning and implications of the responses to each question. The misconceptions identified or hypothesised in this study are viewed as representative of the learning challenges faced by students enrolled for the CHE 2521 module. The concepts with most incorrect responses are deemed as ones for which misconceptions abound. Conversely, those associated with few incorrect responses are viewed as containing less or insignificant level of misconceptions. The author is also aware of the fact that the number of implied misconceptions does not necessarily say anything about the degree to which they impact on a student's learning. The incorrect responses to questions on certain concepts may be few, but of such a serious nature that they hamper learning to a greater degree.

Different approaches for the elimination of misconceptions associated with various chemistry concepts are reported in the literature. It is further confirmed from literature that the methods developed in the framework of the constructivist learning theory are used to remove such misconceptions (Üce and Ceyhan 2019).

4 Results

The results of this study are based on the analysis of students' answers to the diagnostic test. For the sake of clarity, the responses to the test's individual questions are discussed in separate subsections. Representative responses to each of the questions are reported in tables and the inferences on students' preparation are outlined in the same subsection.

4.1 Responses and Misconceptions About Chemical Bonding

Representative responses to questions related to covalent bonding, ionic bonding and intermolecular forces (Question 1 (a) and (b)) are reported in Table 5.1.

The responses show that students have made a distinction between ionic (metals-nonmetals) and covalent bonding (between non-metals) which do not necessarily exist in nature (Nahum et al. 2004). For example, bonding in coordination compounds show characteristics of both ionic and covalent (ibid). There does not seem to be a clear understanding of the role of electrons in bonding. This is in agreement with the findings by Nicoll (2001), who verified that students confuse ionic, covalent and hydrogen bonds and cannot define covalent bond.

The responses also show that students do not understand bonds and intermolecular forces and cannot distinguish bonding the two. Even though they have come across bonding theory and concepts from a 1st year module, they have not been able to construct accurate conceptual frameworks. These results suggest that students passed the previous exams (high school and 1st year) through memorization. They also suggest that students may have passed those prior examinations through other means rather than understanding: they may have copied from other students during the exams or brought illegal materials into the examination rooms.

The literature suggests that misconceptions related to chemical bonding – like those mentioned above – can be eliminated by making students create models and use them while expressing concepts about chemical bonds (Üce 2015) and by

Table 5.1 Representative responses to a question about chemical bonding

***Question**: Distinguish between the following:*

(a) Covalent and ionic bond

(b) Intermolecular forces and bonds

#	**Responses**
1	Covalent bond occurs between non-metals
	Ionic bond occurs between non-metals and metals
2	Intermolecular forces hold electrons in an atom
	Bonds are strong forces of attraction between ionic molecules
3	Intermolecular forces are joining together of two molecular with attractive force
	Bonds are joining together of two molecules
4	Intermolecular forces between electrons which keep the electrons close to each other
	They bind two or more elements without the sharing of electrons
	Bonds are forces that attracts electrons of an atom
5	Intermolecular forces are those that attract electrons in an atom. They occur in the particles of an atom or compound. Intermolecular forces occur when there is a sharing of electrons

preparing guiding materials on ionic bonds based on the constructivist model (Kayalı and Tarhan 2004).

4.2 Responses and Misconceptions About Electronegativity

Representative responses to a question related to electronegativity are reported in Table 5.2.

Students seem to generally understand that electronegativity has to do with ability or force that facilitates the transfer of particles (e.g. electrons). They generally believe that electronegativity has to do with the ability that facilitates transfer of particles between substances. The misconceptions reveal themselves when this 'ability' is expressed in terms of a force, energy or a process. The transfer process refers to 'gaining and attracting' for some students and 'losing and repelling' for others. The particles appear to be electrons to some students and atoms to others. Substances between which particles are transferred are electrons, atoms or between the two (an electron and an atom). Some explanations (e.g., answer 15) have a 'social' connotation; this is an example of what Nahum et al. (2004) refers to as anthropomorphic explanations. Few other students seem to have no clue about what electronegativity is; this is indicated by answers such as 1, 3, 14, etc. Similar to Nicoll (2001)'s subjects, no student linked electronegativity to bonding (bond polarity in particular). However, students' responses were generally not far off from the

Table 5.2 Representative responses to a question about electronegativity

Question: *Discuss electronegativity*	
#	**Responses**
1	Ability of an atom to react and move
2	It gives an indication on weather an element
3	It is the charge that an atom has
4	It is the energy of an atom
5	Ability of electron to attract an electron for itself
6	Ability of an atom to attract a lone pair of electrons
7	Measure of the ability of an atom in a molecule to move from one element to another
8	Measure of attraction of atom to attract another atom
9	A process in which an atom attracts an atom to itself
10	Energy required to remove an electron in an atom
11	Capability of an electron to attract other electrons to itself
12	Force of repulsion due to electrons
13	A number of electrons that are present in the element
14	Ability of an atom to bond
15	Measure of the ability of an atom to compete for an electron
16	Ability of an atom to donate or lose electron

correct answer. Good grasp of electron configuration is essential for understanding periodic properties, e.g. electronegativity of elements.

Demircioğlu et al. (2005) reported the misconceptions regarding trends on the periodic table (for example, those of electronegativity and ionisation energy). They used narration as a remedy for identified misconceptions. The narration scopes include associating chemistry concepts to daily life (e.g. items in supermarket); establishing social and technological structure in chemistry class properly; helping improve students' attitude in the class through presentations of scientific concepts in daily situations; and raising individuals with scientific literacy.

4.3 Responses and Misconceptions About Hybridization

Representative responses to a question about the hybridization of atomic orbitals are reported in Table 5.3.

The hybridization concept seems to be the least understood, and the majority of answers show a wide range of misconceptions coupled with total misunderstanding. It was rarely associated with bond formation, e.g. 'formation of new special bonds', 'overlapping of atoms or orbitals', 'fusion of elements', etc. These definitions suggest lack of understanding of concepts such as atoms, elements and orbitals. Most answers to this question were really off the mark, as indicated by definitions such as 4, 5, and 7 in Table 5.3. Incorrect conceptions and descriptions were varied, and most of them suggested unfamiliarity with general chemistry texts or literature; it is the case of answers 4, 5, 6 and 7 in Table 5.3. Reading the responses one gets the perception that students have not fully engaged with the material related to the concepts. It is as if they are using guess work, like in answer 11.

Table 5.3 Representative responses to a question about hybridization

Question: Define Hybridization	
#	**Responses**
1	Mixing of electrons on the orbitals form other orbitals
2	Mixing of atoms in order to get a more complex molecule
3	Mixing of atomic orbitals to give new atomic orbital
4	Reaction of hydrogen with other molecules, thus hydrogen is added
5	Addition of elements to an atom which makes it to change shape and characteristics and geometry
6	It is the state of the formation of many structures formed as a result of delocalization of the electrons in a molecule
7	It refers to the exchange of an electron from different atom bonded together
8	It refers to the overlapping of orbitals of different atoms
9	It is associated with the formation of new bonds. Bonds between atoms break in order to form *spd* bonds
10	It describes how atom overlap each other and the type of geometry formed
11	Elements fused to form newly hybridized elements

Some students seem to capture information about the concepts in the form of chunks, but do not know how the relevant pieces of information relate to each other; examples are answers 3 and 7. Some of the answers were not far off; for example, answers 1 and 2 and, even closer, answer 3. These answers may suggest that the students have the main idea, but got somewhat confused when expressing it; but they could also indicate lack of understanding of chemistry concepts such as atoms, orbitals, and molecules.

The above results are similar to those obtained by Hanson et al. (2012) when they investigated misconceptions regarding atomic orbitals and hybridization among undergraduate chemistry students. The diagnosed misconceptions emanated from poor conceptualization of the octet rule, the shapes of hybrid orbitals and the driving force of hybridization, among other factors. The implementation of deliberate use of a conceptual teaching approach, in line with the cognitive theory, is recommended, as it has been reported to improve conceptual understanding among undergraduate students.

4.4 Responses and Misconceptions About Resonance

Representative responses to a question related to resonance are reported in Table 5.4.

Some answers were not far off, such as answers 5 and 6; others were outright wrong or showed total confusion of terms, like answers 2 and 3. Again, these responses suggest lack of in-depth understanding of basic chemistry terms such as element, molecules, chemical formula, chemical structure, etc. On the other hand, some responses (including answers 1, 2 and 3) may also suggest that students may have an idea of what resonance is and can perhaps recognize it, but they do not have the appropriate vocabulary to describe it. Answer 4 suggests that students appear to use surface approach to learning and do not engage with concepts deeply enough, and this leads to widespread confusion of terms. One also gets the perception that students rarely consult reading materials or other sources such as textbook and chemistry literature. The description in answer 6 suggests that there is a difference

Table 5.4 Representative responses to a question about resonance

Question: *Define resonance*	
#	**Responses**
1	Resonance shows direction of chemical reaction
2	It is the rearrangement of electrons in atoms
3	Resonance is a substance used to neutralize the charges or to remove charges of chemical solutions
4	The structural molecules which have more than two structural molecules
5	It is the same structural formula but differ in arrangement of elements
6	Different atoms or substances form bonds where the bonds rotate in the molecule or reaction to give different structures

in the way students make meaning from concepts; in other words, students have different frames of reference from those of the teacher, textbooks and literature. Thus, these responses show the need to put more focus on finding out how students make meaning of the chemistry concepts (Nahum et al. 2004).

Furthermore, the reported answers on resonance structures also highlight students' thinking in relation to chemical reactions, electrons and atomic arrangements, charges and bonds. Literature reports misconceptions associated with resonance structures as being related to misconceptions on concepts like reaction equilibrium, charges, stability, atomic and electron arrangement (Widarti et al. 2017). It is recommended that emphasis be placed on drawing Lewis structures, clearly distinguishing between isomers and resonance structures, and depicting resonance structures, in addition to implementing innovative constructivist teaching methods, e.g. using multiple representations, to eradicate identified misconceptions.

4.5 Responses and Misconceptions About the Periodic Table

Representative responses to a question related to the description of groups and periods of the periodic table are reported in Table 5.5.

Groups and periods in the periodic table represent the most basic terms and language used widely in chemistry. Seeing the confusion prevalent around these terms (e.g., answers 1, 2) is sad, because it means that students cannot clearly participate in, or follow some discourses and discussions related to the trends in the periodic table. Views such as those expressed in answers 3 and 1 show that students have no idea what these terms are about. Since students are unsure about these concepts, they may be taking chances by writing as many views as possible (even as contradictory as the views above) with the hope that one of them may be correct. This

Table 5.5 Representative responses to a question about the periodic table

Question: *Describe groups and periods in the Periodic Table*	
#	**Responses**
1	Groups are rows across the periodic table. Vertical elements
2	Periods are columns on the periodic table. Horizontal elements.
3	Periods are vertical elements and are from left to right on the periodic table.
4	Groups of atoms arranged according to specific categories, e.g. metals are grouped together on the periodic table
5	Groups are classified according to stability. They are horizontally arranged and start from top to bottom
6	Groups are read from left to right in the periodic table.
7	Periods are all rows from 1 to 18.
8	Groups represent the outer shell electrons of atoms.
9	Periods represent the distance between the outer shell electrons and the nucleus of atoms.

shows that they are just guessing. Even more confusion is shown by responses 8 and 9. Some responses are not too far off, like answers 4 and 5.

The misconceptions associated with the meaning of groups and periods in the periodic table can be remedied by using analogy, e.g. how items in the supermarket are packed. The knowledge of electron configuration is essential for understanding groups and periods in the periodic table. Misconceptions on this are often revealed by the students' inability to identify the group and period to which a certain element belongs.

4.6 Responses and Misconceptions About the Lewis Structures

For this question, students were required to draw the structures; therefore, very little comments were given. However, the answers (drawings) revealed the following unexpected occurrences:

- Some Lewis structures with three and more atoms contained no central atoms when they should have had it
- Some structures violated the octet rule
- The lone pairs were situated on either side of the central atom and separated into single electrons.

The familiarity with the Lewis structures of molecules is fundamental to understanding covalent bonding. However, the familiarity with Lewis structures among students seems limited; in particular, the understanding of the key concepts to be applied when drawing a Lewis structures seems limited. The errors in the drawings also raise suspicions as to whether students consult reading materials such as textbooks and other sources. Sound knowledge of electronic configurations and the octet rule is essential for mastering Lewis structures. Luxford and Bretz (2014) also identified lack of understanding of the octet rule as a first step towards misconceptions relating to bonding representations like the Lewis structure.

4.7 Responses and Misconceptions About Electron Configuration

Representative responses to a question related to electron configurations are reported in Table 5.6.

The responses show that students associate electron configuration with inappropriate other concepts, e.g. oxidation number (answer 1). This clearly indicates confusion of terms, and limited understanding of basic aspects of atomic structure theory. Therefore, accurate treatment of basic atomic structure theory is necessary to overcome these misconceptions.

Table 5.6 Representative responses to a question about electron configuration

Question: Describe electron configuration	
#	**Responses**
1	It tells us whether an atom is able to lose electrons quicker or easier or if it adds electrons
2	It tells if element has high reactivity, physical properties and stability

Table 5.7 Representative responses to a question about the oxidation number

Question: Describe oxidation number	
#	**Responses**
1	Number of electrons lost
2	A measure of how high is the degree to react with oxygen
3	Number of electrons an atom has in a periodic table
4	Reason in which an atom loses or gain an electron
5	State in which an atom is oxidized
6	Total number of electrons an atoms has
7	Number left when a elements reacts with a oxygen
8	Number that has been reduced
9	The number an atom has in a compound
10	It's the gaining of electrons
11	The number added or donated to a substance
12	The number showing how an atom is oxidized
13	Oxidation number that represents oxidizing agent
14	Loss of electron of an element
15	It is the number presented by irons of an atom

4.8 Responses and Misconceptions About the Oxidation Number

Representative responses to a question related to the oxidation number are reported in Table 5.7.

The answers show a wide range of misconceptions. They give one the idea that students have only superficially engaged with the theory, but do not have a complete grasp of the concept. It seems they only remember some facts about the concepts, but cannot present full and coherent explanations. For instance, they are aware of electron transfer, but without knowing how it happens, and they recall that the oxidation number somehow relates to the number of electrons lost. There seems to also be confusion of terms, as issues totally unrelated to the oxidation number are included in the discussion, like in answers 2 and 3.

Shehu (2015) also identified similar misconceptions with regard to the determination of which species is oxidized or reduced and the determination of the oxidation state of the species involved in a reaction. The diagnosis of misconceptions needs to be followed by the planning of lessons which integrate new information in order to enhance unlearning of alternate conceptions while adopting new ways of thinking; setting up convincing laboratory experiences, and using more structural models or technology–based method, are recommended as remediation methods.

4.9 Responses and Misconceptions About the Octet Rule

Representative responses to a question related to the octet rule are reported in Table 5.8.

Students seem to have a general idea that the octet rule involves eight electrons, stability and bonding. However, they do not seem to know how the three features relate, as clearly highlighted by answers 2, 3, and others. Students seem to have a tough time linking terms/features that are essential in the octet rule. It is as if they remember only certain words and terms that are disjointed, likely as a result of memorization.

Misconceptions about the octet rule are found also in other contexts. For instance, Ozmen (2004) reports about students not linking the term octet to the number 8, as some students involved in his study believed that 'nitrogen atom can share five bonding pairs of electrons', leading to ten electrons around nitrogen atom instead of eight.

5 Conclusions

A qualitative research method using a diagnostic test as a tool of analysis was applied to test the level of preparedness and the presence of misconceptions for students enrolled for the 2nd year Inorganic Chemistry module (course code: CHE

Table 5.8 Representative responses to a question about the octet rule

***Question**: Describe the octet rule*	
#	**Responses**
1	Shows that an atom doesn't not need to show bonds because it is already stable.
2	It's the rule that mainly includes compounds with eight electrons
3	It requires eight bonds for stability
4	It occurs when there are eight electrons or bonds
5	The chemical substance must have maximum electrons of eight
6	It is the molecular which contacts the bonding of eight electrons
7	States that each atom involve in the bond formation must have a total of eight electrons at the end of a reaction.
8	Maximum of sharing eight electrons in a molecules

2521) at the University of Venda. The results revealed that misconceptions related to hybridization, electronegativity, octet rule, chemical bonding, groups and periods in periodic table, electronic configuration, Lewis structures, oxidation number and resonance abound. The concepts tested form part of high school and 1st year syllabi. In this way, students do not seem to be adequately prepared for the concepts in the 2nd year level module.

Students seem not to have formed and developed appropriate frames of references needed to deal with chemistry concepts. These are students who have successfully completed their 1st year general chemistry module, in which concepts included in the diagnostic test form part to the syllabus. There are reasons to believe that student engage in practices that make them effective at test-taking and examination-writing, without understanding the subject matter.

Close analysis of students' responses to the diagnostic test leads to a number of conclusions.

The students' responses suggest that surface learning is the common approach among the respondents. This leads to memorization of facts. Memorization without fully understanding the core meaning of concepts leads to forgetfulness, recalling of disjointed facts, and at times total confusion. The responses to most questions showed that students remember only part of the information and often fail to link or associate facts accurately, as the responses were largely incoherent and at times confusing.

The lack of adequate chemistry vocabulary is also apparent. This may be a direct result of not acquiring textbooks or not consulting other chemistry sources. It is further confirmed by the difficulty students have in engaging and discussing chemistry concepts. They are unable to use disciplinary terminology in their discussion. One gets the impression that students have a lack for words or vocabulary to express themselves. Therefore, end up using wrong words, thus failing to convey the message they intend to. Language proficiency also seems to be lacking as one reads more and more badly constructed sentences.

Close analysis of CHE 2521 promotion results for 2013, 2014 and 2015 showed pass rates over 50%. The fact that the majority of students end up passing the CHE 2521 module suggests something related to teaching approaches or assessment criteria. There may be misalignment between learning outcomes and assessment criteria (Gibbs 1999). It seems as if the level at which students are assessed (in tests, assignments and examinations) is lower than expected. The assessment may be testing recall, by requiring short or one word answer. This sort of assessment does not test true understanding of concepts. Students look to assessment criteria to determine what they need to pass the course (Gibbs 1999). If the assessment seeks for conceptual understanding by requiring students to make their thinking and understanding explicit, then the quality of learning may improve.

Appendix: The Diagnostic Test

CHE 2521(Diagnostic test; Inorganic chemistry)				
Instructions				
1. Answer all questions giving as much information as you can.				
2. Write neatly and legibly.				
3. The test will not count for marks.				
Questions				
1. Distinguish between the following: (a) Covalent and ionic bond (b) Intermolecular forces and bonds				
2. Discuss electronegativity and how it changes in the periodic Table.				
3. Define hybridization.				
4. Define resonance.				
5. Describe features of the Periodic Table: (a) periods (b) columns				
6. Draw the Lewis structure for the following compounds:				
(a) CO_3^{2-}	(b) NO_3^-	(c) XeF_6	(d) ICl_2^-	(e) BF_3
7. (a) Describe the electronic configuration of a substance. (b) What information can one gather from electronic configuration of a substance?				
8. Describe oxidation number.				
9. Describe the octet rule.				

References

Barke, H. D., et al. (2009). *Misconceptions in chemistry, Chapter 2* (pp. 21–35). Berlin/Heidelberg: Springer.

Biggs, J. (1999). What the student does: Teaching for enhanced learning. *Higher Education Research & Development, 18*(1), 57–75.

Boo, H. K. (1998). Students' understandings of chemical bonds and the energetics of chemical reactions. *Journal of Research in Science Teaching, 35*(5), 569–581.

Demircioğlu, G., Ayas, A., & Demircioğlu, H. (2005). Conceptual change achieved through a new teaching program on acids and bases. *Chemistry Education Research and Practice, 6*(1), 36–51.

Emondson, K. M. (2005). In J. J. Mintzes, J. H. Wandersee, & J. D. Novak (Eds.), *Assessing science understanding – A human constructivist view* (pp. 19–36). London: Elsevier Academic Press.

Fischer, G. L., & Scott, I. (2011, October). The Role of Higher Education in Closing the Skills Gap in South Africa. In *The World Bank Human Development Group, Africa Region, Background Paper 3: The role of Education*, pp. 1–53.

Gibbs, G. (1999). Using assessment strategically to change the way students learn. In S. Brown & A. Glasner (Eds.), *Assessment matters in higher education: Choosing and using diverse approaches*. Maidenhead: SRHE/Open University Press.

Gilbert, J. K., & Zylberstajn, A. (1985). A conceptual framework for science education. The case study of force and movement. *European Journal of Science Education, 7*, 107–120.
Hanson, R., Sam, A., & Antwi, V. (2012). Misconceptions of undergraduate chemistry teachers about hybridisation. *African Journal of Educational Studies in Mathematics and Sciences, 10*, 45–54.
Hasan, S., Bagayoko, D., & Kelley, E. L. (1999). Misconceptions and the Certainty of Response Index (CRI). *Physics Education*, 294–299.
Kayalı, H., & Tarhan, L. (2004). A guiding material implementation based on the constructivist-active learning to eliminate misconceptions on ionic bonds. *Hacettepe University, Faculty of Education Journal, 27*, 145–154.
Kind, V. (2004). *Beyond appearances: Students' misconceptions about basic chemical ideas* (pp. 1–78). Institute of Education, University of London.
Luxford, C. J., & Bretz, S. L. (2014). Development of the bonding representations inventory to identify student misconceptions about covalent and ionic bonding representations. *Journal of Chemical Education, 91*(3), 312–320.
Nahum, T. L., Hofstein, A., Mamlok-Naaman, R., & Bar-Dov, Z. (2004). Can final examinations amplify students' misconceptions in chemistry. *Chemistry Education Research and Practice, 5*(3), 301–325.
National Research Council. (1997). *Science teaching reconsidered: A handbook*. Washington, DC: The National Academies Press. https://doi.org/10.17226/5287.
Neumann, R. (2001). Disciplinary differences and university teaching. *Studies in Higher Education, 26*(2), 135–146.
Nicoll, G. (2001). A report of undergraduates' bonding misconceptions. *International Journal of Science Education, 23*, 707–730.
Ozmen, H. (2004). Some student misconceptions in chemistry: A literature review of chemical bonding. *Journal of Science Education and Technology, 13*(2), 147–159.
Palmer, D. (1999). Exploring to link between students' scientific and non-scientific conceptions. *Science Education, 83*, 639–659.
Palmer, D. (2001). Students' alternative conceptions and Scientifically acceptable conceptions about gravity. *International Journal of Science Education, 23*, 691–706.
Schoon, J. K., & Boone, J. W. (1998). Self-efficacy and alternative conceptions of science of pre-service elementary teachers. *Science Education, 82*, 553–568.
Shehu, G. (2015). Two ideas of redox reaction: Misconceptions and their challenges in chemistry education. *IOSR Journal of Research & Method in Education (IOSR-JRME), 5*(1), 15–20.
Taber, K. (2000). Chemistry lessons for universities?: A review of constructivist ideas. *University Chemistry Education, 4*, 63–72.
Tan, K. C. D., Goh, N. G., & Treagust, D. F. (2002). Development and application of a two-tier multiple choice diagnostic instrument to assess high school students' understanding of inorganic chemistry qualitative analysis. *Journal of Research in Science Teaching, 39*(4), 283–301.
Üce, M. (2015). Constructing models in teaching of chemical bonds, ionic bond, covalent bond, double and triple bonds, hydrogen bond and molecular geometry. *Educational Research and Reviews, 10*, 491–500.
Üce, M., & Ceyhan, I. (2019). Misconception in chemistry education and practices to eliminate them: Literature analysis. *Journal of Education and Training Studies, 7*(3), 202–208.
Widarti, H. R., Retnosari, R., & Marfu'ah, S. (2017, August). Misconception of pre-service chemistry teachers about the concept of resonances in organic chemistry course. *AIP Conference Proceedings, 1868*, 030014-1–030014-10. https://doi.org/10.1063/1.4995113.
Zoller, U. (1990). Students' misunderstandings and misconceptions in college freshman chemistry (general and organic). *Journal of Research in Science Teaching, 27*(10), 1053–1065.

Chapter 6
Are the Newly Formed Kenyan Universities Ready to Teach Externally Examined Diploma Courses in Analytical Chemistry?

Warren A. Andayi

1 Introduction

In the last 5 years, Kenya has experienced massive horizontal expansion of university education, with an increase of public universities from 7 to more than 30 (Government of Kenya Report 2015; KUCCPS 2020). Middle level diploma colleges were acquired to accommodate the new universities colleges (NUCs), and this has compromised the quality of STEM education, despite increased funding and perceived improvement in staff competency via an injection of highly qualified teaching staff into these former technical colleges. As a transition phase, the new institutions have continued to offer the diploma courses while they prepare to launch undergraduate degrees, as a source of income, or to retain the relevant former technical college staff. Furthermore, STEM diploma courses are still offered in the new universities to ensure undisrupted supply of middle level technical manpower. Most of these diploma courses are externally examined by the Kenya National Examination Council (KNEC).

Diploma training falls under the Technical, Industrial and Vocational Training (TIVET) sections of universities or independent technical institutions. The TIVET education is meant for participants to acquire practical skills, knowhow, and understanding which are necessary for employment in a particular occupation, trade or group of occupations (Atchoerene and Delluc 2001). TIVET qualifications (usually up diploma and certificate levels) were traditionally offered by technical schools and national polytechnics; thus, the takeover by NUCs can be regarded as one by inexperienced hands. The takeover included infrastructure, human resources (including most of the academic staff), academic programs, and students. A question arose on whether the new "tenant" is competent in handling the dual responsibilities of running the diploma and the degree programmes in chemistry. Although

W. A. Andayi (✉)
Murang'a University of Technology , Murang'a, Kenya

L. Mammino, J. Apotheker (eds.), *Research in Chemistry Education*,
https://doi.org/10.1007/978-3-030-59882-2_6

degree programmes are fast displacing diploma courses from the NCUs, they both can still be accommodated if proper requirements are taken into account and modifications are put in place (Fig. 6.1).

Analytical chemists are trained in a range of methods to enable them investigate the chemical nature of a substance, which basically involves identifying and understanding a substance and its behaviour in different conditions. Analytical chemists can work in areas as diverse as drug formulation and development, chemical or forensic analysis, process development, product validation, quality control and toxicology. This diversity of opportunities implies that a chemist needs good social and technical skills, and this implies the need for high quality training skills.

The teaching of externally examined diploma courses in analytical chemistry is experiencing a number of challenges in the NUCs. A preliminary study was done to establish the readiness of newly created universities to teach an externally (KNEC) examined diploma in analytical chemistry. It was hypothesized that the NUCs readiness or ability to teach this course is on the decline or is inadequate. Readiness or lack of readiness were investigated with respect to the following aspects: staff competency and academic freedom; diversity among the academic staff; adequateness and appropriateness of infrastructure; and rising cost of training. To investigate these challenges, preliminary studies were done on three NUCs, with supplementary data from secondary sources i.e. literature. The direct data collection was carried out through the following approaches: interviewing of lecturer, administrators and students; observations; focus group discussions involving lecturers and students; and ethnography. The secondary data sources included students' academic records from within the NUCs, government reports and relevant publications.

2 Results

This section reports the results obtained, organising them according to the terms of reference and criteria listed in the previous section. These terms and criteria will constitute the titles of the subsections. Illustrative data are reported in tables,

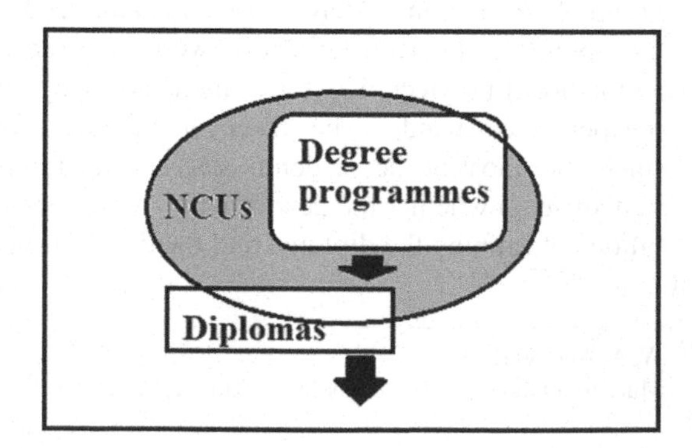

Fig. 6.1 Simultaneous accommodation of degree programmes and diploma courses in NCUs

selecting three NCUs (respectively denoted as NCU-1, NCU-2 and NCU-3) and a long established university (denoted as EU).

2.1 *Diversity Among the Academic Staff*

The American Council on Education (2012) recognises collective diversity among institutions as one of the great strengths of America's higher education system and states that it has helped make it the best in the world. Diversity is essential in enabling educational institutions to serve the needs of a democratic society and of the increasingly global scope of the economy. Therefore, diversity in the student bodies, faculties, and staff is important for the provision of high-quality education. The accrued benefits to higher education include: enriched educational experiences, strong communities and workplace, enhanced personal growth and healthy society, and enhanced economic competitiveness (Fig. 6.2). Therefore persons trained in NCUs with adequate diversity will be well prepared for the workplace and will overall lead to economic growth. NCUS with good diversity profiles stand a chance of being more competitive and attractive to talent than those without. Perfect diversity environment enhances teamwork since there is lesser overbearing dominance of one group.

Some of the NUCs involved in the study are located in rural areas and cannot easily attract highly or adequately qualified manpower; thus, merit and diversity are not key criteria in their academic staff recruitment processes. This is further exacerbated by the rapid expansion in the student population (Table 6.1), warranting desperate need for more academic staff irrespective of qualification. In all the three NUCs considered as samples, there was minimal academic staff diversity in terms of ethnicity. However, good diversity was found in terms of age and academic qualifications of the academic staff (Table 6.1). The benefits deriving from age and academic diversity seem to be negated by the absence of ethnic diversity, resulting in observed limited co-operation and team work among the staff. As an aftermath of the effects of the 2007/08 post-election violence in Kenya on national integration

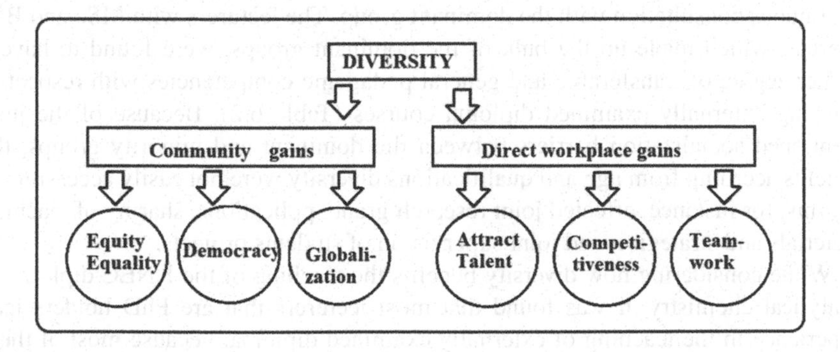

Fig. 6.2 Some of the possible benefits of diversity in an academic workforce

Table 6.1 Demographics, location type and exam performance in the selected universities

Institution	Student population % growth	Setup	Lecturers' population	% Change in dip exam mean grade	% of lecturers from indigenous community	% of lecturers with PhD
NCU-1	400	Rural	>100	−12	50	52
NCU-2	460	Rural	>100	−3	71	44
NCU-3	200	Urban	<100	0.0	35	50
EU	10	Urban	>500	N/A	No dominant group	80

and cohesion, it appears that threads of subtle animosity and mistrust among staff members from different ethnic backgrounds are still present in some of the NCUs under study.

Among the NUCs considered, it was explicit that indigenous people of the location of the NUCs constitute one or two dominant groups (Table 6.1). As expected, people from same ethnicity tend to relate more easily and amicably amongst themselves than with persons from different ethnicities. In all NUCs studied, the dominant groups had the habit of communicating using their native language (instead of English or Swahili) even in official functions, disregarding the presence of other staff members who do not have a clue about the language. A language barrier was deliberately created against minority groups, which caused the latter not to relate closely with the ethnic majority. In such a situation, the productivity synergy expected to accrue from age and qualifications diversity is negated. It was observed that lecturers from minority groups tended to integrate easily amongst themselves for professional and social reasons, and hence this group can be said to have higher level of cross-cultural competency. Some staff members preferred consulting within minorities' group than outside – with the dominant groups. The lecturers from minority groups tended to be more highly qualified (having MSc and PhD qualifications) than the indigenous staff who, in most cases, were former employees of the technical institutions prior to conversion to a NUC. Unfortunately, most of these highly qualified people had little experience in teaching Diploma students (the minimum qualification for a technical college teacher is bachelors, compared to the NCUs' Masters' requirement). Therefore, the less experienced minorities needed continuous consultation with the dominant group. The lecturers with MSc and BSc degrees, which made up the bulk of the dominant groups, were found to have a higher degree of transferable and general pedagogic competencies with respect to teaching externally examined diploma courses (Table 6.2). Because of the just-mentioned socialization barriers between the dominant and minority groups, the benefits accruing from age and qualifications diversity were not easily accessible to all; this, for instance, affected joint research grant applications, sharing of teaching materials and strategies, and joint supervision of students projects.

While considering how diversity benefits the teaching of the KNEC diploma in analytical chemistry, it was found that most lecturers that are PhD holders lack experience in the teaching of externally examined diploma, because most of them had recently graduated or were teaching at mainstream universities before joining

Table 6.2 Chemistry department structures and academic staff work-experience in the selected universities

Institution	% of lecturers with experience in teaching KNEC diploma candidates	Chemistry department merged with other sciences	Adequacy of library space and IT infrastructure as a percentage	% of lecturers with adequate training in instrumentation (>7 analytical instrument)	% of lecturers who spend >24 h per week on research & extension
NCU-1	50	Yes	40	50	14
NCU-2	60	Yes	50	30	10
NCU-3	63	Yes	50	56	10
EU	10	No	75	100	70

NCUs. They did have good hands-on and theoretical grounding in chemical experimentation and instrumentation (Tables 6.2 and 6.3). On the other side, most lecturers without PhDs tended to have invaluable teaching experience and excellent pass track records in KNEC analytical chemistry, which is an asset for institutions offering this course. The non-PhD holders stated that they were competent enough when teaching diploma courses and thus did not need to consult PhD holders, and they considered the latter to be inexperienced in teaching diploma courses. If integration and cohesion are achieved, then the diversity will definitely enhance quality teaching, as the academic staff experiences will be synergised to train high quality analytical chemistry technicians. More synergy from diversity was observed in NUCs near big cities than those in the rural (Table 6.4). Synergistic diversity was expressed in the form of sharing of teaching materials and insights, examinations preparation, laboratory experimentation/practical work, course allocation and time table preparation, content depth and difficulty etc.

2.2 *Rising Costs of Training*

The unit cost of using PhD holders to teach diploma students was found to be high when compared to teaching the same students using lecturers with MSc and BSc qualifications. Generally, the academic staffs with PhDs earn more than colleagues with lower qualifications doing the same tasks. This is even made more costly as pass rates in the terminal diploma exams are continuously declining, and most of the PhD holders are rarely engaged in research and consultancy due to absence of requisite infrastructure and funding support in NCUs (Tables 6.1 and 6.2).

A drop in KNEC analytical chemistry diploma exam performance was reported in all the NUCs under consideration. This drop was attributed to the replacement of the pedagogically competent MSc/MEd and BSc holder teachers with PhD holders. It would be economically sensible to train more B.Com students than Diploma analytical chemists using PhD holders.

Table 6.3 Teaching strategies, facilities and academic freedom in the selected universities

Institution	% of lecturer who prefer demonstrations over hands on practical/ experiments	% of lecturers who assign >6 lab practicals/ experiments to diploma students per term	% of lecturers who feel they have academic freedom when involved in teaching of diploma students	% of PhD holders who prefer teaching KNEC diploma classes	Institutions with designated/ labelled labs	% of lecturers above 50 years of age
NCU-1	80	0	0	20	Yes	10
NCU-2	90	0	0	22	Yes	0
NCU-3	88	0	0	33	Yes	6
EU	N/A	N/A	N/A	N/A	No	50

Table 6.4 Academic staff social interactions and exams preparation in the selected universities

Institution	% of lecturers that do revision	% of lecturers that complete the syllabus	% of students who feel adequately prepared for examinations	% of indigenous lecturers that cooperate with non-indigenous	% of non-indigenous lectures that feel to be fully involved or consulted on key depart/Sch./faculty decisions
NCU-1	55	55	45	10	20
NCU-2	68	70	60	33	30
NCU-3	78	81	50	25	21
EU	N/A	N/A	N/A	50	N/A

Most analytical chemistry diploma classes were found to comprise less than 10 students, which further illustrates the training of this group to be uneconomical. Some remedial strategies like the introduction of outreach and marketing initiatives has not helped improve the enrolment into Diploma courses in analytical chemistry. The marketers usually go out advertising part-time and fulltime undergraduate courses for the popular streams (business and IT), with little or no mention of KNEC examined STEM diplomas. This negatively influences the pre-college pipeline of KNEC diploma in analytical chemistry and results in dwindling enrolment in the same.

Resource allocation in NCUs is biased towards undergraduate laboratories and resources. The driving force is to ensure standard undergraduate facilities for a competitive edge in attracting government sponsored undergraduate students, since this translates into more funding allocation from the government. By 2015, the government started sponsoring diploma students via TIVETA (Technical and Industrial Vocational Training Agency). However, in most NCUs the financial contribution of

TIVETA sponsored students is meagre relative to what the government gives towards undergraduates and postgraduates in public universities. Therefore, most of the NUCs have deliberately expanded degree enrolment and shrunk resources for diploma courses. Enrolment in the NUCs analytical diploma classes shrunk from classes with over 20 students to less than 5 students in less than 3 years of their establishment.

2.3 Staff Competency and Academic Freedom

Due to the issues of perceived incompetency of PhD holders to teach KNEC diploma in analytical chemistry, whenever there was a choice between an undergraduate course and a diploma course, the PhD holder would prefer teaching the undergraduate course. Most PhD holders felt that teaching diploma was cumbersome and tedious, and robbed them of the academic freedom of deciding what to teach and how to examine it (Table 6.3). This is because KNEC diploma courses have prescribed course outlines and the diploma awarding examinations comes long after the teaching (after 3 years of teaching). Academic competency and freedom, as expected in the university, thus become an issue of concern. Academic freedom is the belief that the freedom of inquiry by faculty members is essential to the mission of the academy and the principles of academia. Thus, in essence, the scholars should have freedom to teach or communicate ideas or facts without risking repression, joblessness or any negative consequence (Bankauskiene 2005). Without experience in teaching diploma courses, some lecturers' expectations of freedom of academic discourse have been aborted and their creativity limited.

With staff scarcity, the NCUs continue to hire people whose first degree is not closely related to their 2nd and 3rd degrees, resulting in a half-baked academic staff. Given that most chemistry departments are merged with physics and biological sciences into one department (Table 6.2), some lecturers are forced to take up courses that do not fall within their specialization. For instance, a chemistry lecturer can be forced to teach the biochemistry components of the diploma in analytical chemistry. This is unacceptable because such a lecturer did not specialize in biochemistry and his/her approach will be less biochemical and more of a biological-organic chemistry nature. It is well known that having a major in the subject taught has a significant positive impact on students' achievements, and that a link exists between teachers' undergraduate majors and students' achievements in mathematics and science (Kuenzi 2008; Allen 2003). This can be a factor contributing towards the mediocre performance of Diploma studies in the NUCs (Table 6.1).

Designating and labelling separate lecture rooms/buildings for undergraduate and diploma students was reported in two of the NUCs considered, and this can be interpreted as academic discrimination which goes against the spirit of university education (Table 6.3). Such a practice sets diploma students apart, hinders integration, socialization, sharing of knowledge and demoralises them from pursuing undergraduate studies. It also fuels animosity and superiority/inferiority complexes amongst students pursuing chemistry at diploma and degree levels. Furthermore this negates the benefits accruable from diversity.

The technical staff competency was under question since the PhD holders expected independent operation of the technical staff in terms of selection and implementation of students' experiments; however, this was not the case, since lab technicians were used to receiving express instructions on the schedule of the experiments to be performed. Such conflict of expectations affects the practical training of the diploma students and, due to time constraints, some experimental work is excluded from the training program, above all when the teacher is a PhD holder (Tables 6.2 and 6.3). This fact has been noted elsewhere in the 2015 Government report, where it is stated that the curriculum for TVET remains outdated even though technology changes. In view of this, curricula need to be reviewed by persons in the industry rather than by incompetent trainers. The problem is further exacerbated by the shortage of trainers, coupled with the inadequate professional training of those who are expected to teach at this level (Government of Kenya Report 2015).

2.4 Facilities and Equipment

TVET is constrained by limited infrastructure and obsolete equipment (Government of Kenya Report 2015). To enable Technical universities to carry out their mandate, there is need for a deliberate and continuous government funding of infrastructure, specifically for science and technology laboratories. However, this is not happening according to what is needed (Table 6.2). This has therefore fuelled competition for scarce lab space/personnel between undergraduates and diploma students and, since the latter are outnumbered and contribute less to the university coffers, they ultimately get sidelined and are given less laboratory time and attention. Indeed, direct funding for the improvement of physical infrastructure is less frequent in conventional fully fledged universities worldwide, and this problem is worsened in NUCs by the practice of converting technical colleges to universities. In one of the NUCs, a block that had been built solely to host science labs was converted into computer labs; this was done to help an institution that once hosted 90% science and engineering students to accommodate over 85% business and IT students in less than 3 years.

Most of the NUCs considered in this study had labs divided into two sections, physical sciences and biological sciences, with the latter hosting chemistry and physics. The KNEC analytical chemistry diploma exams do not have a laboratory practical section; hence, for most NUCs it has never been a priority to equip the chemistry labs with the instruments required for diploma analytical chemistry training. On the basis of the KNEC diploma and the American Chemical Society (ACS) recommendations, most laboratories do not meet the minimum training requirements. Instrumentations in the key categories of chromatography, electrochemistry, optical, atomic and molecular spectroscopy, as well as some wet chemistry apparatus are lacking. It has to be noted that the laboratories in the NUCs considered in this study had some of the requisite equipment, such as UV-VIS, preparative TLC tanks/columns and analytical TLC plates, extractions apparatus, electrochemical cells and pH meters, but these remained inadequate and underutilised. Access to sources of secondary chemical data was limited; e.g. access to FT-IR, NMR, MS and HPLC data sheets or plots was limited due to lack of access to chemistry journals or due to the incompetence or lack of experience on the side of lecturers (Table 6.4). This gap was usually assumed to be to some extent compensated by the internship attachment, where the intern is advised to seek attachment to industrial, research and university laboratories. Internships thus were the only recourse to covering the hands-on education gap in the diploma training. A lot is expected to be learned during internship because, prior to it, most learners do not have basic chemical analysis skills like sample collection, sample handling, sample weighing, or sample storage. Despite this, the majority of the teachers felt that their lab facilities were adequate for the diploma teaching with respect to KNEC diploma examinations requirements.

3 Mitigation Strategies

To mitigate the worrying trends described in the previous section, the government has mandated the NCUs to mentor and supervise the construction or upgrading of new technical colleges within their vicinities. This is a deliberate move to create replacement of technical schools that were swallowed by the NCUs. It can be compared to removing fruiting plants and replacing them with seedlings, and it essentially implies reversed development. Once the mentored institutions/new technical colleges become operational, they are expected to take over the TIVET mandate from the NUCs, albeit as novices.

It is recommended that a regional centre/laboratory of excellence in analytical sciences is established in order to enable quality education of analytical chemistry technicians and effective channelling of resources for the achievement of intended goals in training essential manpower, as well as in developing and disseminating best practices in chemical education. Even though this is a new concept in Kenya, it stands to enhance the quality of training offered to analytical chemistry technicians.

The number of student teachers involved in technical education courses is not increasing because of shrinking job openings, since most leading technical colleges have been converted to NUCs. In the long run, the ability of newly created universities to teach KNEC examined diploma may come to naught.

To enhance the training of diploma holder technicians then there must be prior increase in the supply of new STEM teachers and improvement of the skills of the current STEM teachers (Kuenzi 2008). The reverse of this has happened for the diplomas in analytical chemistry and other STEM courses examined by KNEC. With emphasis on degrees in NCUs, the former diploma teachers' skills are rendered into disuse as most are forced to teach undergraduate courses and pursue graduate studies (MSc and PhD) in order to remain relevant. All this may not be lost if hybrid university courses, incorporating both sound theoretical grounding and practical/technical skills, are adopted as it is done globally in Technical universities, with notable examples in Germany and South Africa. So far, the Technical University of Mombasa and the Technical University of Kenya are deliberately and explicitly pursuing these models, but most NCUs that sprouted from technical colleges have not embraced this model except in their names. In the government's budgetary allocation to universities, only 10% go to the hiring of staff and to infrastructure in NCUCs (Nganga 2011); this too needs to be increased.

To mitigate the issue of competing interests, most NCUs have a TIVET sections as a government requirement and, in addition, the NUCs are mentoring new technical institutions in their vicinities, to grow more opportunities for the training of diploma technicians. Furthermore, the government has set aside funds for the construction of physical amenities and the purchase of equipment, and for setting up eight national polytechnics nationwide. This is in tandem with one of the aims of TIVET education in Kenya, that is, to mobilize resources to rehabilitate facilities in public TIVET institutions so as to ensure quality training (Nyerere 2009). However, this aim has not been achieved in most NUCs. Even though NCUs continue training chemical analysis technicians, there is still need for continual improvement of the quality of staff and infrastructure, so as to close the gaps.

References

Allen, M. B. (2003, July). *Eight questions on teacher preparation: What does the research say?* Education Commission of the States.

American Council on Education. (2012). *On the importance of diversity in higher education*. Retrieved on June 20, 2016 from http://www.acenet.edu/news-room/Documents/BoardDiversityStatement-June2012.pdf.

Atchoerene, D., & Delluc, A. (2001). *Revisiting technical and vocational education in sub-saharan Africa, an update on trends, innovation and challenges*. Paris.

Bankauskiene, N. (2005). *The expression of teacher competencies in action research field; the case based study of KTU teacher education program "pedagogy"*. Presented at the European conference on educational research, University of Dublin.

Government of Kenya. (2015, Novemebr). *Education sector report 2016/17 – 2018/19 MTEF budget report.*

Kenya Universities and colleges Central Placement Service KUCCPS. (2020). *List of accredited colleges and universities*. https://kuccps.net.

Kuenzi Jeffrey, J. (2008). *Science Technology, Engineering and Mathematics (STEM) education background, federal policy and legislative action* (Congressional Research Service Reports. Paper 35).

Nganga, G. (2011). *Kenya: Double student intake kicks off at Kenyatta*. University World News. Retrieved June 26, 2016 from http://www.universitynew.com/article.php?story.
Nyerere, J. (2009). *Technical and vocational education and training [TiVET] sector mapping in Kenya*. Amersfoort: Edukans Foundation.

Chapter 7
"Closing the Circle" in Student Assessment and Learning

Francis Burns

1 Background

American higher education is quite different from the European higher education (Theilen 2011; Huisman and van Vught 2009). The American model incorporates a liberal arts core, including of mix of science, history, English, and social science. The first and second years tend to have fewer degree-specific courses. An American chemistry student would take only one chemistry course during his or her first year. In contrast, European chemistry students typically focus on their degree. A European chemistry student would take a series of chemistry courses, such as first-year analytical chemistry, first-year physical chemistry, and first-year organic chemistry. Both continents produce excellent university graduates, but there are significant differences in the degree programs.

A large proportion of American chemical education research has focused on the pedagogy and learning associated with general chemistry. Why? General chemistry serves as a "gate-way" course for the vast majority of science, engineering, and medical degrees at almost all American universities. Few students become scientists, engineers, or doctors without success in general chemistry. As a result, lecture tends to be filled with a large and diverse student population. These students tend to be first-year students, but more advanced students will also enrol in general chemistry. American students have varied interests, career goals, as well as different levels of prior knowledge and skills. The diversity of students increases a teacher's challenges associated with selecting effective pedagogical methods. General chemistry covers content from first-year physical, inorganic, and analytical chemistry. General chemistry is usually split into two one-semester courses. General chemistry I (CHEM 121 at Ferris State University) covers the first-half of the content. Most students take this course during their fall semester in their first-year of university.

F. Burns (✉)
University of South Carolina Salkehatchie, Allendale, SC, USA
e-mail: fmburns@mailbox.sc.edu; francisburns473@gmail.com

L. Mammino, J. Apotheker (eds.), *Research in Chemistry Education*,
https://doi.org/10.1007/978-3-030-59882-2_7

General Chemistry II (CHEM 122 at Ferris State University) covers the second-half of the content. Most students take this course during their winter semester in their first year of university.

One confounding variable in American chemical education research is the semester cohort system. Due to the importance of general chemistry, the course is often repeated during an academic year with different cohorts of students. The author has observed anecdotally that fall semester cohorts and spring semester cohorts are not equivalent. Winter semester cohorts have fewer students enrolled in the Honors Program. Many of the winter semester students failed chemistry during the prior fall semester. Many of the winter semester students needed to complete additional mathematics as a prerequisite, which caused them to lag behind their peers. However, the author has observed that winter cohorts appear to have the same expectations for success as fall cohorts.

The author's pedagogical approach described in this chapter illustrates an observation made by Carl Rogers: "…the curious paradox is that when I accept myself just as I am, then I change (Rogers 1961)." In the author's experience as a professor, students often do not know that they have a problem until late in a course, and then they often do not know that solutions exist for their problems. Students will generally do whatever they need to do in order to succeed, if they understand what they need to do and the rationale behind these actions.

Why is "self-evaluation" important? A person's ability to evaluate himself or herself has been related to overall performance level (Dunning et al. 2003). The author's personal experience has shown that students are particular poor at self-evaluation. Students cannot correct problems that they do not recognize. Instructors are frequently more aware of common problems than students, but the majority of the power to mitigate these problems lies with students.

2 Variables Affecting Student Performance

The literature has investigated many variables proposed to affect student performance. Some of these variables are pedagogical in nature, such as developing new instructional and assessment methods. Other variables can be characterized as more fundamental, such as the effect of multitasking in learning and student performance (Bowman et al. 2010; Zhang 2015; Sana et al. 2013). The author assigns an end-of-semester writing assignment in which current students write a letter to future CHEM 121 students in order to recommend activities to be done and activities to be avoided. Based upon a cursory review of these student letters, students endorse four variables as particularly important: study skills, time management, anxiety management, and attentional control.

Study skills have long been recognized as essential for consistent, high-achieving student performance (Dendato and Diener 1986; Robbins et al. 2004). Robbins and his co-authors performed a meta-analysis in order to identify factors affecting college student outcomes. The authors analyzed 36 studies that examined the relationship between academic-related skills (study skills) and student success and retention. They found that effective study skills have a strong positive relationship with student success and retention.

During fall semester of 2010, the author offered to schedule fifteen minute conferences with his chemistry students. All students were recommended to take advantage of the opportunity, but these conferences were strongly recommended for students earning a "D" or "F" on the first test. Approximately 30 students out of 120 students discussed their progress with the author. Due to ad-hoc nature of information gathering, one cannot draw firm conclusions from these conferences. However, the author observed that struggling students tended to limit their normal out-of-class activities to the required homework activities. They also tended to defer their learning activities until shortly before tests. One failing student was bewildered by the author's assertion that she needed to spend at least two hours per day to "rescue" her grade. She truly did not know how to spend her time productively beyond a simple review of lecture notes prior to a test. In contrast, the "B" and "A" students employed more learning activities on a regular basis. Their test preparation focused on review of material, not learning material.

In addition to study skills, time management techniques have often been taught at American universities. One research group developed a multidimensional questionnaire in order to assess four factors affecting time management: setting goals and priorities, mechanics of planning and scheduling, perceived control of time, and preference for disorganization (Macan et al. 1990). The study also assessed students' stress level and grade point average through self-report. Students' perceived control of time had the greatest correlation between stress, satisfaction, and performance.

Anxiety is an omnipresent factor in life, but it can have positive effects. For example, the author's youngest son did very little job hunting during his final quarter at the University of Chicago. Why? He felt little anxiety or "inner pressure" to engage in a difficult activity. Once anxiety developed after graduation, he started to look for a job. Test anxiety has a similar effect on students. Too little or too much test anxiety tends to produce sub-optimal results. Dendato and Diener investigated the effects of cognitive/relaxation therapy, study skills training, and combined therapy/training for students suffering from severe test anxiety (Dendato and Diener 1986). Relaxation/cognitive training was found to reduce anxiety but failed to improve test scores. Study skills training failed to reduce anxiety or improve test scores. However, the combined use of relaxation therapy and cognitive skills training significantly reduced anxiety levels and increased test scores.

Attentional control or the ability to focus on tasks appears to be inversely related to multitasking behavior. Sana and others determined that students multitasking on a laptop during lecture scored lower on a test compared with students who did not multitask (Sana et al. 2013). Another study broadened multitasking to students' overall usage of information and communication technologies (Junco and Cotten 2012). The authors found a negative correlation between students' use of information and communication technologies and their academic performance.

3 Technologies Used for Teaching and Learning

Modern educational practices have developed an array of technologies to teach and learn chemistry. Some practices have been used since the dawn of chemistry or earlier, such as the Socratic Method or practicals. Other practices became commonplace with the advent of computers and the Internet, such as online homework. Chemical educators must select their tools carefully. They need to consider the desired learning outcomes, as well as the constraints faced by their students. The present work utilized a variety of tools to promote teaching and learning.

3.1 Writing Assignments

Berthoff discusses the value of reflection through the use of "dialectical notebooks," which are informal tools for the recording of ideas, questions, passages from books, and most importantly, reflections upon them (Berthoff 1987). Both science and science education courses have made extensive use of student writing for purposes of student reflection. For example, Grumbacher found that "learning logs" were an effective tool for high school physics. She stated that her students' logs improved their problem solving ability, integrated experience and theory, and improved her students' enjoyment of physics (Grumbacher 1987). Indeed, Byers emphasized reflection as a major task for his chemistry students, in an effort to develop their independent learning skills (Byers 2007).

3.2 Online Homework

Online content-based homework has become a common tool for chemistry teachers. This type of homework encourages students to practice problem solving and answer conceptual questions. Most software packages will provide additional assistance or direct references to a textbook, if a student submits an incorrect answer. Instructors also receive valuable information about their students' learning, which can guide instruction. The author used a particular online homework package associated with the textbook "Mastering Chemistry" (Pearson Higher Education 2013). The web-based software collects a range of information: points earned by students, the time spent on assignments, and other information, such as commonly submitted wrong answers. All of this information can be used by instructors for assessment purposes.

3.3 Classroom Polling Devices

The use of classroom polling devices or clickers has been extensively reported in the literature. MacArthur and Jones reviewed 92 reports applicable to the use of clickers in chemistry classrooms. The devices permit rapid collection of student answers to questions posed in lecture for formative evaluation. Clickers have also been used to collect student answers for quizzes and tests (MacArthur and Jones 2008). Researchers have used clickers to collect survey data: communications (Bunz 2005) and psychology (Langley et al. 2007). They directly compared clickers to other standard methods used for surveys and determined that the device was a suitable alternative.

3.4 Course Percentage/Grades

Course percentage and grades provide a global measure of student performance in chemistry courses, including laboratory and homework. Many variables affect grades. Their relative influence can change from professor to professor and semester to semester. This limits the utility of course grades as assessment instruments. Nonetheless, the author uses course grades as one measure of students' performance.

3.5 Tests

In contrast to course grades, standardized tests measure students' performance in a limited set of concepts and skills, but do so in a fashion that permits comparison across institutions, professors, and semesters. This project uses two standardized tests to assess student performance: *California Chemistry Diagnostic Test* serves as a pretest (University of California – Berkeley 2006) and *First Term General Chemistry* test serves as a post-test.

In summary, chemical educators have long been concerned with student learning. The chemical education community has invested considerable resources into the development of effective instructional activities and technologies, such as guided inquiry methods and online homework. The education community has also developed many qualitative and quantitative instruments in order to measure student learning. Figure 7.1 illustrates effective student assessment as a triangular model (Carnegie Mellon University 2015). Although "learning objectives" must be student-centric, the Carnegie Mellon University model focuses more upon teaching and assessment activities, which are teacher-centric. The author would like to

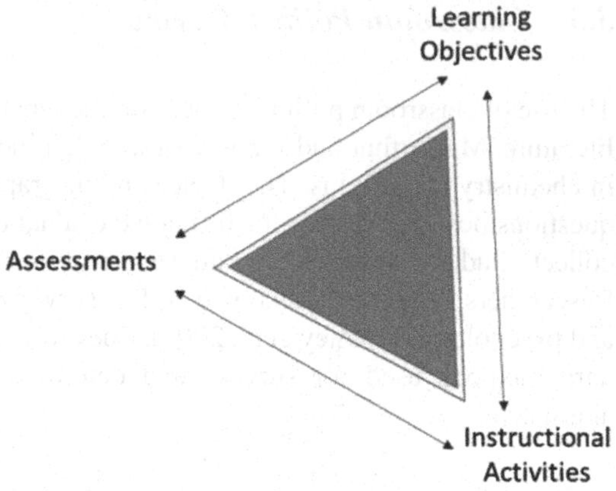

Fig. 7.1 Relationships (Alignments) between learning objectives, instructional activities and assessments (Carnegie Mellon University 2015)

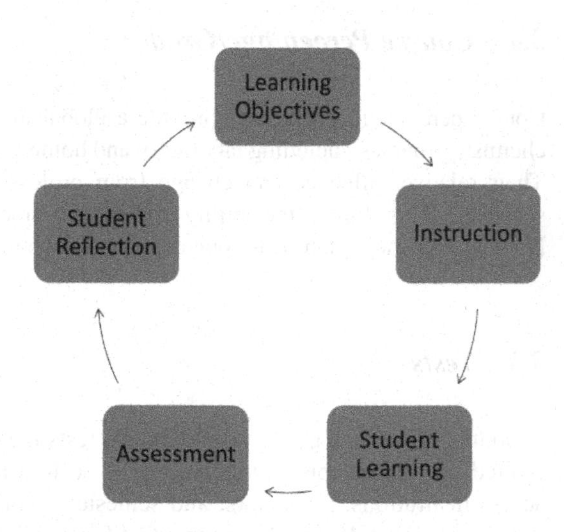

Fig. 7.2 "Closing the Circle" in teaching and learning by expanding the relationships between teacher-centric activities and student-centric activities

propose an alternative model for learning, which is more student-centric (Fig. 7.2). Students are the individuals with the greatest capacity to make improvements in their learning processes. They enter their course of study with expectations of success. In the author's experience, most students will exert themselves in their studies only as needed.

Unfortunately, a student's perception of his or her success can be quite inaccurate until too much time has passed, resulting in the failure of multiple chemistry tests. If students received sufficient accurate information about their performance at the beginning of a particular course or degree program, they will recognize their strengths and weaknesses, and then address them. Suppose a student recognizes that she lacks adequate time management skills due to a self-assessment worksheet (College of Retention and Student Success, Ferris State University Seminar). As a result, she would likely suffer a lower grade due to inadequate time management; then this student would probably become interested in learning about effective time management strategies.

Learning is an inherently individual process. Ann Berthoff stated succinctly the nature of thinking, which is an essential process in learning: "Thinking begins with perception: all knowledge is mediated (Berthoff 1987)." New knowledge is constructed from a base of old knowledge and perceptions. As a student acquires new information, he or she needs to integrate it into the framework, modifying the framework as needed. This process is inherently reflective. A student can superficially engage in the educational process (i.e., come to lecture, complete homework questions, and read the textbook – maybe even ask questions) and still not achieve satisfactory outcomes.

4 Methods

The author taught CHEM 121 students in the spring of 2014 at Ferris State University (Michigan, USA). He found the course to be quite challenging. 35.8% of the students did not pass the course. He discussed this result with his colleagues, who told him that this was typical for the spring semester course. The author defines "failing" as earning a D, F or W (Withdrew from the course). The fall semester 2013 CHEM 121 course had a very different result: only 11.8% of the students did not pass the course. The author thought that there were interventions that could be employed to lower the failure rate for spring semester courses of CHEM 121 to make the failing rate comparable to fall semester courses of CHEM 121.

The foundation for the methods described in the following paragraphs resides in a specific student assessment activity. Student grades from all spring semester courses in CHEM 121 from 2007 to 2014 (N = 936) are summarized in Fig. 7.3. Four different instructors taught during this time period, who used very different pedagogies. 27.5% of all of the students earned a D, F, or W, which will hereafter be labelled as "DFW grades." The percentage of DFW grades assigned in a particular semester ranged from 17.5% to 35.8%. Based upon historical evidence, one could reasonably expect that a little more than a quarter of future cohorts of students would fail to earn an adequate grade for their degree program (C or better).

The author has taught CHEM 121 several times at Ferris State University, but three cohorts seemed most suited for comparison: spring semester 2014, fall semester 2014, and spring semester 2015. The pedagogical methods described hereafter were implemented during the spring semester of 2015. Spring semester 2014 and fall semester 2014 cohorts served as benchmarks for assessing the spring semester 2015 cohort.

Ferris State University attempts to prepare its students for the rigors of university learning. All first-year students are required to take a freshman seminar: Ferris State University Seminar (FSUS). FSUS addresses issues related to learning, such as time management and study skills. Unfortunately, many disparaging comments were heard from first-semester chemistry students, indicating that many of them are not receptive to changing their habits. It would not be surprising if a sizable percentage of the second-semester CHEM 121 students believed that time management and study skills were not important during their first semester.

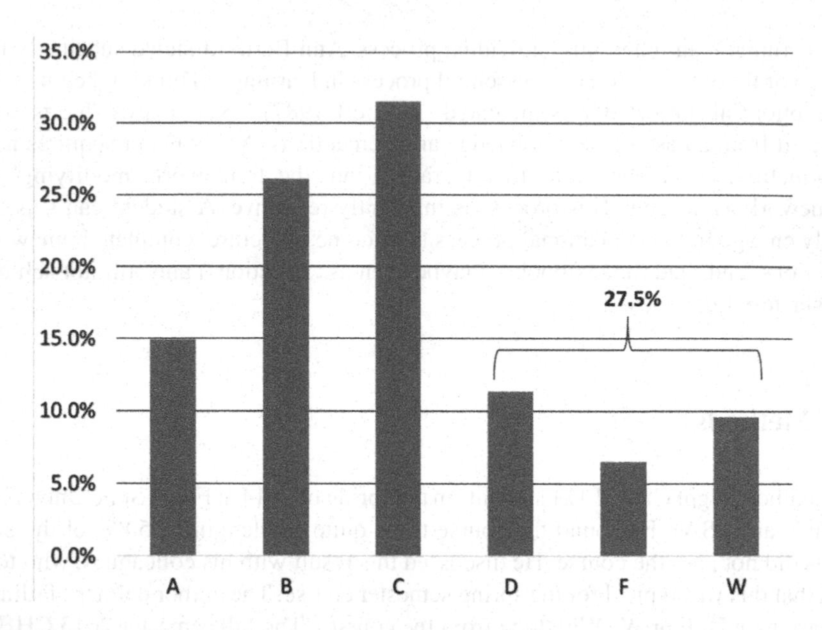

Fig. 7.3 The grade distribution for students enrolled in the spring semester sessions of General Chemistry I (CHEM 121) from 2007 through 2014 (N = 936)

The author decided to improve his students' receptivity to learning new ideas. He implemented his pedagogical plan during spring semester CHEM 121 course in 2014 (N = 93 students). The students completed their *first* 3-h practical in a classroom instead of a laboratory. There were four groups of 24 or fewer students, which permitted the use a workshop model. Each session started with a frank discussion about past student performance. The students were shown the historical trends in student grades for spring semester CHEM 121 courses. The lecturer conveyed the gravity of the students' situation by telling them "Based upon prior semesters, 27.5% of you have already 'washed out' of CHEM 121. Look around the room. Four or five of you have already failed." Almost every student would look shocked. The "prediction" was followed with a very simple statement "We can do better."

The workshop portion started after this short presentation. Each student completed the "Procrastination Quotient" worksheet (College of Retention and Student Success, Ferris State University Seminar), which was not collected by me. I presented FSUS material for time management, and then students were required to schedule time for studying. Subsequently, FSUS material related to study skills (College of Retention and Student Success, Ferris State University Seminar Program) was presented. The learning outcome for the workshop can be summarized in the following terms: (1) encourage students to accept themselves as they are, and then (2) provide students with the tools needed to change.

During the semester, the author used a variety of formative and summative assessment instruments to provide students with the information that they need to improve their self-awareness. The first week of CHEM 121 was used to administer two assessment instruments: (1) a standardized chemistry pre-test (University of California – Berkeley 2006), and (2) a "multitasking assessment" instrument, which was based on an instrument provided by FSUS. Too many of the spring semester 2014 students engaged in multi tasking with their laptop or mobile device during lecture. The activity required students to measure the time required to complete a simple task while focusing solely upon the task, and then repeating the task while multitasking in a manner similar to driving a car and using a mobile to text a friend. After collecting the multitasking data from the spring 2015 students, the data from the fall 2014 cohort were displayed for their reflection (Fig. 7.4). Multitasking slows down task completion: 15.9 seconds for a simple task. Similarly, texting or chatting during lecture impedes a student's ability to learn chemistry. As a result, multitasking has negative consequences for time management and learning.

Classroom polling devices, online homework (Master Chemistry), and semester tests for instructional and assessment purposes were also used. Students' data were summarized in tables or charts, and then the summarized results were shared with the class during lecture. Students were encouraged to compare their individual performance with peers and were also provided suggestions for improving learning and testing. For example, students' answers on their first weekly quiz were collected using a classroom polling device (clicker). At the same time, students were asked to self-report the time spent on studying chemistry. The data were summarized and

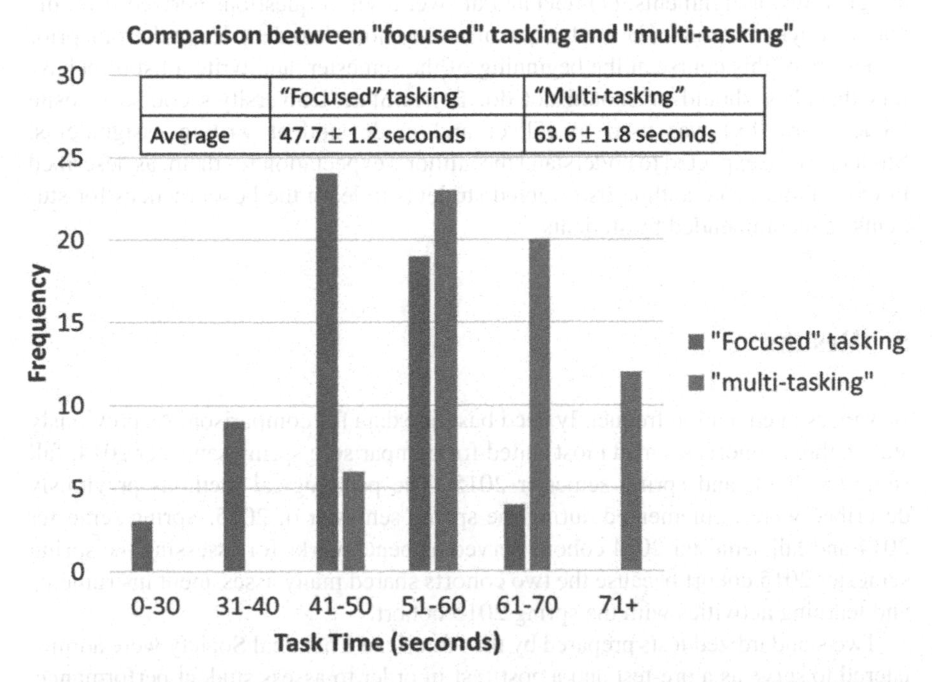

Fig. 7.4 The experimental results of focused tasking compared with multitasking for the fall 2014 cohort. There is a significant difference in the average task time required by students ($p = 0.05$)

Table 7.1 Summary of student scores on the first weekly quiz: Overall class average (78.9%), as well as the average student scores based upon the self-reported time spent on studying chemistry

Overall average	78.9%
Average time spent by students earning a 55% or lower	3 hours
Average time spent by students earning a 78% or higher	7 hours

Table 7.2 Descriptive statistics for the California Chemistry Diagnostic Test using normative data. The groups had no significant difference with respect to ANOVA ($p > 0.05$)

	Average ± standard error	Median
Spring 2014 (N = 129)	30.4% ± 2.2%	20.0%
Fall 2014 (N = 100)	25.5% ± 2.2%	20.0%
Spring 2015 (N = 88)	29.4% ± 2.4%	25.0%

discussed at the next lecture (Table 7.1). Finally, the *First Term General Chemistry* was administered as the course's final examination, which is commonly used as an assessment tool (American Chemical Society, Division of Chemical Education, Examination Institute 2005).

Writing assignments provide students with an opportunity to engage in guided reflection. This tool was used at different stages during a semester. Students were assigned two assignments: (1) read and answer a set of questions derived from the course's syllabus, and (2) read a set of anonymous "student letters" from prior semesters of this course at the beginning of the semester, and write a list of behaviors that they should or should not do. Ferris State University's course website (Blackboard 9.x) was used to collect and grade student writing assignments. Students were expected to understand the author's expectation for them, as described in the syllabus. The author also wanted students to learn the best practices for students as recommended by students.

5 Results

Advances in education frequently need baseline data for comparison. As previously stated, three cohorts seemed most suited for comparison: spring semester 2014, fall semester 2014, and spring semester 2015. The pedagogical methods previously described were implemented during the spring semester of 2015. Spring semester 2014 and fall semester 2014 cohorts served as benchmarks for assessing the spring semester 2015 cohort because the two cohorts shared many assessment instruments and learning activities with the spring 2015 cohort.

Two standardized tests prepared by the American Chemical Society were administered to serve as a pre-test and a post-test in order to assess student performance. The *California Chemistry Diagnostic Test* served as a pre-test over three semesters: Spring 2014, Fall 2014, and Spring 2015 (Table 7.2). Similarly, the *First Term General Chemistry* served as the post-test over the same semesters (Table 7.3). The

Table 7.3 Descriptive statistics for the First Term General Chemistry using normative data. All of the groups had significant differences with respect to ANOVA ($p < 0.05$)

	Average ± standard error	Median
Spring 2014 (N = 120)	57.5% ± 2.3%	58.0%
Fall 2014 (N = 101)	48.5% ± 2.5%	52.0%
Spring 2015 (N = 87)	41.9% ± 2.4%	39.0%

Table 7.4 Descriptive statistics for the Overall Course Percentages for each semester. The groups had no significant difference with respect to ANOVA ($p = 0.05$)

	Average ± standard error	Median
Spring 2014	79.2% ± 1.1%	78.4%
Fall 2014	79.2% ± 1.4%	80.2%
Spring 2015	80.0% ± 1.1%	79.7%

students' raw score was converted into a percentile rank using the tests' normative data. The different cohort's results are summarized in Table 7.3.

The cohorts are fairly matched, based upon the analysis of variance (ANOVA) of the students' pre-tests. The chemistry content covered each semester is consistent, but the learning activities vary. As a result, it is not surprising that student performance varies across the semesters based upon ANOVA. Every student received an overall percentage based upon their performance on different activities and assessments: practicals, quizzes, homework, writing assignments, class participation, semester tests, and the post-test (Table 7.4). Despite differences in the semesters' results for the post-test, there was no significant difference in the overall percentage which correlate to students' final grades (A, B, C, D, or F). This result was found to be surprising because a semester was not finished with a predetermined course average in mind. In addition, some variation occurs between semesters with regards to the basis for calculating the overall percentage. For example, semester tests were provided 45% of the overall percentage in spring 2014, but 50% in fall 2015.

Learning activities varied across semesters, especially writing activities and practicals. Weekly quizzes were an important part of the course because they supplied evaluation data to the instructor; they provided 10% of the overall course grade. It was also hoped that students would also use their personal results as a measure of their performance. The points varied across semesters, but the same technology was used: classroom polling devices. The quizzes were administered at the beginning of class, and then the content was reviewed immediately. There is no significant difference in student performance based upon weekly quizzes, which are a measure of their efforts to stay up with the content covered in lecture (Table 7.5). However, the first three weeks of lecture show a very different result (Table 7.6). The spring 2014 cohort's performance was much poorer when compared to the other two cohorts. All of the cohorts diminished in number between the first semester examination and the end of the semester, but the spring 2014 cohort lost the greatest number.

Mastering Chemistry homework shows a similar result to the weekly quizzes. All of the cohorts diminished in number between the first semester examination and the

Table 7.5 Descriptive statistics for the Overall Weekly Quiz scores for each semester. The groups had no significant difference with respect to ANOVA (p = 0.05)

	Possible points	Average ± standard error	Median
Spring 2014 (N = 119)	54	64.1% ± 1.6%	63.0%
Fall 2014 (N = 103)	104	65.3% ± 1.5%	65.0%
Spring 2015 (N = 91)	99	63.5% ± 1.0%	62.6%

Table 7.6 Descriptive statistics for the Overall Weekly Quiz scores for the period prior to the first semester test. Spring 2014 was significantly different from the other two semesters with respect to ANOVA (p = 0.05)

	Possible points	Average ± standard error	Median
Spring 2014 (N = 134)	18	54.7% ± 1.8%	55.6%
Fall 2014 (N = 110)	24	66.1% ± 1.9%	66.7%
Spring 2015 (N = 92)	27	69.2% ± 1.7%	70.4%

Table 7.7 Descriptive statistics for the Overall Mastering Chemistry scores for each semester. The groups had no significant difference with respect to ANOVA (p = 0.05)

	Possible points	Average ± standard error	Median
Spring 2014 (N = 119)	515	79.2% ± 1.1%	78.4%
Fall 2014 (N = 103)	346	79.2% ± 1.4%	80.2%
Spring 2015 (N = 92)	308	80.0% ± 1.1%	79.7%

Table 7.8 Descriptive statistics for the Overall Mastering Chemistry scores for the period prior to the first semester test. All of the cohorts were significantly different from with respect to ANOVA (p = 0.05)

	Possible points	Average ± standard error	Median
Spring 2014 (N = 133)	172	82.5% ± 1.9%	88.1%
Fall 2014 (N = 110)	80	93.9% ± 2.9%	109.2%
Spring 2015 (N = 93)	80	110.7% ± 2.3%	118.7%

Note: students could earn bonus points for homework

end of the semester, but the spring 2014 cohort lost the greatest number. There are no significant differences in cohorts' average homework scores at the end of the semester (Table 7.7). In contrast, the first three weeks of work shows considerable difference between the cohorts (Table 7.8). The spring 2014 lagged behind the other cohorts with their completion of Mastering Chemistry, which provided online practice of content and problem-solving skills. Surprisingly, the spring 2015 cohort far exceeded the performance of the fall 2014 cohort. 48% of the fall 2014 cohort consisted of students enrolled in the Honors Program. These students were pursuing pharmacy or other medical degrees, and they are typically the best prepared and motivated students at Ferris State University. The author did not determine how many of the spring 2015 cohort were enrolled in the Honors Program, but he doubts if the percentage approached the number for fall cohort. It should be note that the

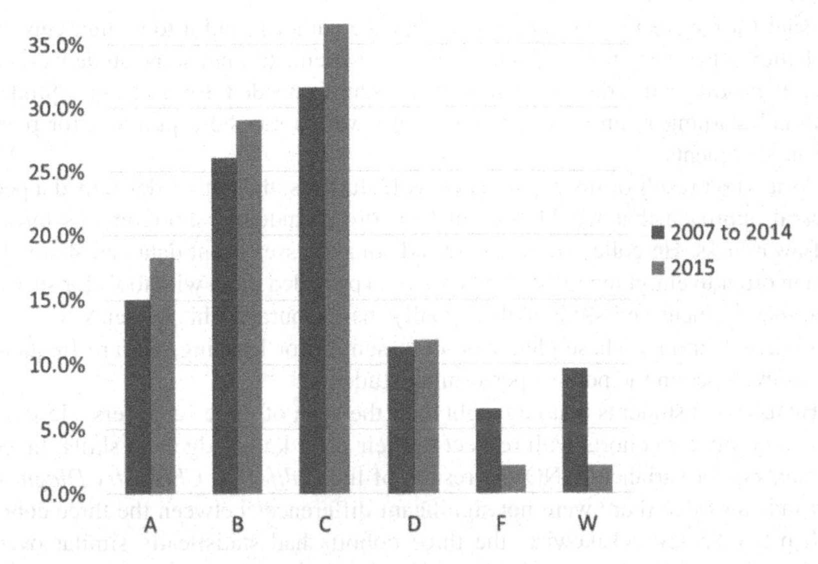

Fig. 7.5 Historical trends in the final course grades in spring CHEM 121 courses (2007 through 2014) in comparison with final course grades for spring 2015

author has had a long practice of offering his students both bonus points and penalty for online homework. Online homework provided 10% of the overall course grade. The penalty occurs when a student misses a deadline for homework submission (20% deduction). The bonus occurs when a student fulfils a deadline with a score of 90% or better. A few bonus points are also offered for the "orientation" assignment to encourage students to begin their online homework.

When the spring 2015 cohort began the semester, as described above, the author shocked the students with a very challenging statistic. Historically, 27.5% of students enrolled in the spring CHEM 121course receive a D, F or W grade. The author challenged the students to change this situation. He also provided them with methods to learn chemistry, as well as skills needed to learn, such as time management and focusing on a single task at a time. He maintained pressure on his students by providing timely assessment data, such as weekly quiz scores, homework scores, and periodic tests. He had originally planned to formally require his students to writing assignments to encourage reflection, but he was unable to do this. Never-the-less, the spring 2015 cohort dramatically decreased the percentage of F and W grades (Fig. 7.5). The cohort's overall percentage of DFW grades was 16.1%.

6 Discussion

When the author submitted his grades for CHEM 121 for the spring 2014, he found that many students were going to need to retake CHEM 121 or change their degree program: 35.8% of his students received a D, F, or W grade. This outcome was not unusual for the spring semester course, but the author found it to be unacceptable. He believes that very few students want to fail chemistry, but some students do not have the skills, attitudes or even self-awareness needed for success. "Student-centred" learning requires students to learn, which can be a problem for poorly prepared students.

As a direct result of his experiences with students, the author developed a pedagogical approach that would attempt to address students' shortcomings through self-awareness. He collected and analysed student assessment data and shared this data in order to enlighten his students. He also provided them with tools for success: time management and study skills. Finally, he encouraged his students' to reflect upon their learning. These elements are essential for learning, but are frequently underdeveloped in the poorest performing students.

He assessed students that he taught over the span of three semesters. He started with very similar cohorts with respect to their prior knowledge and skills. Indeed, the analysis of variance (ANOVA) results of the *California Chemistry Diagnostic Test* indicates that there were not significant differences between the three cohorts with $p > 0.05$ level. Likewise, the three cohorts had statistically similar overall course percentages. Based upon these two measures, one could conclude that the three groups are very similar.

For the spring 2015 cohort, the author used instructional time to collect and/or present assessment data, as well as teaching time management and study skills. The students worked harder as measured by Mastering Chemistry and weekly quiz scores. The improved scores directly raised student grades, permitting a larger percentage of students to pass chemistry. The author had assumed that greater effort by students would also positively affect their acquisition of content knowledge and skills. However, the ANOVA results for *First Term General Chemistry* indicate that the cohort for spring semester 2015 did not achieve the same performance level as the other two semesters. This is a curious result that may be explained by a larger proportion of poorer performing students remaining in the course to take the post-test *First Term General Chemistry* assessment. The author's assessment of the three cohorts indicates that student success is much less dependent upon prior chemistry knowledge and skills than their attitude, learning skills, and personal management skills. Interventions that target these student attributes may increase student success more than a strict focus on content delivery.

Acknowledgements This research was supported by the Physical Sciences Department of Ferris State University.

The presentation at the 2nd African Conference on Research in Chemical Education was supported by the following organizations:

- Diné College (Arizona USA), which provided me with travel funds and release time from my instructional duties
- American Chemical Society – Division of Chemical Education, which awarded me a Passer Travel GrantI would like to recognize the support of many mentors and colleagues, especially Dr. Jon Kirchhoff and Dr. Dean Giolando at the University of Toledo, as well as Dr. David Frank at Ferris State University.

References

American Chemical Society, Division of Chemical Education, Examination Institute. (2005). *First term general chemistry* [Measurement instrument]. Milwaukee: American Chemical Society, Division of Chemical Education, Examination Institute.

Berthoff, A. E. (1987). In T. Fulwiler (Ed.), *Dialectical notebooks and the audit of meaning*. Portsmouth: Boynton/Cook Publishers.

Bowman, L. L., Waite, B. M., & Gendron, M. (2010). Can students really multitask? An experimental study of instant messaging while reading. *Computers & Education, 54*, 927–931.

Bunz, U. (2005). Using scantron versus an audience response system for survey research: Does methodology matter when measuring computer-mediated communication competence? *Computers in Human Behavior, 21*, 343–359. https://doi.org/10.1016/j.chb.2004.02.009.

Byers, W. (2007, August). Developing independent learners in chemistry: Promoting a knowledge-based economy. In *41st IUPAC World Chemistry Congress*. Turin.

Carnegie Mellon University. (2015). *Why should assessments, learning objectives, and instructional strategies be aligned?* Retrieved September 19, 2015, from Eberly Center for Teaching Excellence & Educational Innovation: https://www.cmu.edu/teaching/assessment/basics/alignment.html

College of Retention and Student Success, Ferris State University Seminar. (n.d.). *Time management resources – Procrastination quotient* (A self assessment worksheet). Retrieved May 5, 2016, from Ferris State University: http://ferris.edu/HTMLS/colleges/university/fsus/Faculty/topics/TimeMana.htm

College of Retention and Student Success, Ferris State University Seminar Program. (n.d.). *Study skills and learning strategies*. Retrieved May 25, 2016, from Ferris State University: http://ferris.edu/HTMLS/colleges/university/fsus/Faculty/topics/learning.htm

Dendato, K., & Diener, D. (1986). Effectiveness of cognitive/relaxation therapy and study-skills training in reducing self-reported anxiety and improving the academic performance in test-anxious students. *Journal of Counseling Psychology, 33*(2), 131–135.

Dunning, D., Johnson, K., Ehrlinger, J., & Kruger, J. (2003). Why people fail to recognize their own incompetence. *Current Directions in Psychological Science, 12*, 83–87. https://doi.org/10.1111/1467-8721.01235.

Grumbacher, J. (1987). In T. Fulwiler (Ed.), *How writing helps physics students become better problem solvers*. Portsmouth: Boynton/Cook Publishers.

Huisman, J., & van Vught, F. (2009). Diversity in European higher education: Historical trends and current policies. In F. van Vught (Ed.), *Mapping the higher education landscape: Towards an European classification of higher education* (Vol. 28, p. 17). Springer.

Junco, R., & Cotten, S. R. (2012). No A4 U: The relationship between multitasking and academic performance. *Computers & Education*, 505–514. https://doi.org/10.1016/j.compedu.2011.12.023.

Langley, M. M., Cleary, A. M., & Kostic, B. N. (2007). On the use of wireless response systems in experimental psychology: Implications for the behavioral researcher. *Behavior Research Methods, 39*(4), 816–823.

Macan, T. H., Shahani, C., Dipboye, R. L., & Philips, A. P. (1990). College students' time management: Correlations with academic performance and stress. *Journal of Educational Psychology, 82*(4), 760–768.

MacArthur, J. R., & Jones, L. L. (2008). A review of literature reports of clickers applicable to college chemistry classrooms. *Chemistry Education Research and Practice, 9*, 187–195. https://doi.org/10.1039/B812407H.

Pearson Higher Education. (2013). *Mastering chemistry* [online homework for Chemistry: The Central Science, 12th Edition by Brown, T. E.; LeMay, H. E.; Bursten, B. E.; Murphy, C.; Woodward, P.]. Retrieved from http://www.pearsonmylabandmastering.com/northamerica/masteringchemistry/

Robbins, S. B., Le, H., Davis, D., Lauver, K., & Langley, R. (2004). Do psychosocial and study skill factors predict college outcomes? A meta-analysis. *Psychological Bulletin, 130*(2), 261–288.
Rogers, C. R. (1961). *On becoming a person*. Boston: Houghton Mifflin Company.
Sana, F., Weston, T., & Cepeda, N. J. (2013). Laptop multitasking hinders classroom learning for both users and nearby peers. *Computers & Education, 62*, 24–31.
Theilen, J. R. (2011). *A history of American higher education* (2nd ed.). Baltimore: John Hopkins University Press.
University of California – Berkeley. (2006). *California chemistry diagnostic test* [Measurement instrument]. Milwaukee: American Chemical Society, Division of Chemical Education, Examination Institute.
Zhang, W. (2015). Learning variables, in-class laptop multitasking, and academic performance: A path analysis. *Computers & Education, 36*, 82–88.

Chapter 8
Interpretation and Translation of Chemistry Representations by Grade 11 Pupils in the Chipata District, Zambia

Lawrence Nyirenda and Salia M. Lwenje

1 Introduction

The aim of this study was to assess the interpretation and translation of chemistry representations by Chemistry and Science Chemistry pupils at senior secondary schools in Zambia. Chemistry representations are the different forms in which the language of chemistry can be expressed. Johnstone identified three types of chemistry representations: macro, symbolic and micro or sub-micro. He illustrated these using the corners of a triangle (Johnstone 1991, 2006). Over the years, the meaning and interpretation of the corners of Johnstone's triangle has been the subject of many discussions (Taber 2013; Talanquer 2011) and hence we briefly explain the interpretation used in this study. Macro representations are typical of experimental or laboratory activities from which observations are made and conclusions drawn. Chemical symbols, chemical formulae and chemical equations make up the symbolic representations of chemistry. Descriptions of the structure of matter and the changes it undergoes in terms of atoms, molecules or ions constitute the sub-micro representations of chemistry. This interpretation of the chemistry representations is in line with Johnstone's original ideas (Johnstone 1991, 2000, 2006).

To interpret a chemistry representation is to attach appropriate meaning to it. To translate a chemistry representation is transforming it into another form with an equivalent meaning.

Johnstone's triangle, sometimes referred to as the 'chemistry triplet' (Gilbert and Treagust 2009; Talanquer 2011), has influenced a lot of research in chemical

L. Nyirenda (✉)
Chipata College of Education, Chipata, Zambia
e-mail: lawnyirenda1@gmail.com

S. M. Lwenje (Deceased)
Chemistry Department, School of Mathematics and Natural Sciences, Copperbelt University, Kitwe, Zambia
e-mail: salia.lwenje@cbu.ac.zm

L. Mammino, J. Apotheker (eds.), *Research in Chemistry Education*,
https://doi.org/10.1007/978-3-030-59882-2_8

education in the past two to three decades (Bradley 2014; Davidowitz et al. 2010; Gabel 1999; Johnstone 1991; Joseph 2011; Kozma and Russell 1997; Nyachwaya and Wood 2014; Talanquer 2011; Treagust et al. 2003; Wu et al. 2001). It has been noted that conceptual understanding of chemistry involves representing and interpreting chemical problems using the three levels of the chemistry triplet (Gabel 1999). Treagust et al. (2003) reported that pupils were able to understand the how and why of chemistry concepts by linking the macro, submicro and symbolic levels of chemical representations. Davidowitz and her colleagues also found that when students used multiple representations they performed better. They suggested that to help students make links between the different levels they should be introduced to multiple representations including particulate diagrams and physical models (Davidowitz et al. 2010). Joseph (2011) found that pupils' poor understanding of the concepts of chemistry could be attributed to teachers' limited conception of the three levels of chemical representations and a lack of effective teaching methods to facilitate learning. Bradley (2014), has proposed that Johnstone's triangle can be viewed as a closed-cluster concept map that can be used to illustrate inter-relations between macro, sub-micro and symbolic levels.

In Zambian senior secondary schools, two different chemistry syllabuses are offered. These are Chemistry 5070 and Science 5124 (Curriculum Development Centre 2013a, b). Chemistry 5070 is generally referred to as 'pure' chemistry and is normally offered to pupils who are deemed to be academically stronger. The Science 5124, referred to as Science Chemistry, is offered to weaker pupils. This dual approach of teaching chemistry in Zambian secondary schools has been in place since 1983, when the taking of science courses in the form of physics courses and chemistry courses was made compulsory. According to the Science 5124 and Chemistry 5070 syllabuses, successful learners in Zambian secondary schools are expected to easily and effectively switch from one form of chemistry representation to the other. This is deduced from the expected general outcomes reflected in both syllabuses. In either case, Chemistry and Science Chemistry pupils are expected to have a sufficient or adequate understanding of chemistry. However, at the time of this study there was no evidence of any study having been done to verify pupils understanding of chemistry in terms of its three representations. This study sought to obtain baseline data on interpretation and translation of chemistry representations by attempting to answer the following research questions:

1. Are the Grade 11 Chemistry and Science Chemistry pupils able to interpret some concepts in each of the chemistry representations and switch from one form to another?
2. What strategies do teachers use to teach each of the chemistry representations?
3. How often is each of the three chemistry representations used in the teaching of chemistry and science chemistry?
4. To what extent are physical models used in teaching of chemistry and Science Chemistry?

This study was based on Jerome Bruner's theories of constructivism and instructions (Bruner 1977). The three chemistry representations are more or less in line with Bruner's theories.

2 Methodology

2.1 Sampling

The study targeted populations consisting of all Science Chemistry and Chemistry grade 11 pupils and their teachers, in all the 15 schools located in the Chipata district, in the Eastern Province of Zambia. Five schools from these 15 were chosen to participate in the study. The main selection criterion was that the school had at least one class of Chemistry 5070. All the schools in the district had one or more classes of Science 5124. If a selected school had more than one class of either Chemistry 5070 or Science 5124, one class in each category was selected to participate in the study using simple random sampling. In all the schools, the Chemistry 5070 classes had fewer numbers of pupils than the Science 5124 classes. All Chemistry 5070 and Science 5124 teachers in the selected schools participated in the study. Some teachers taught either Chemistry 5070 or Science 5124 only, whilst others taught both groups. Table 8.1 gives the names of the selected schools, the number of pupils in each selected class, and the number of teachers that participated in the study. The sample size was 128 Science 5124 pupils, 58 Chemistry 5070 pupils and 13 teachers. Five pupils from each class, giving a total of 10 pupils, participated in the focus group discussions at each school.

The study was done in November 2014, by which time the grade 11 syllabus had been covered. Permission to conduct the study was obtained from the Head of each selected school. The purpose of the study was explained to each participant. Further, the participants were assured of confidentiality of all information obtained and that participation in the study was voluntary. Participants were informed that there was no payment for participation in the study, but that their participation would contribute to the knowledge of chemistry education in Zambia. Utmost care was paid to ensure that activities ended in a manner that left respondents in a safe emotional state.

Table 8.1 Names of schools and number of pupils and teachers that participated in the study

	Number of pupils		Number of teachers		
Name of school	Sci. 5124	Chem 5070	Sci. 5124	Chem 5070	Sci. 5124 & Chem 5070
Hill Side Girls	20	11	2	1	1
Chipata Day	25	9	1	1	1
Anoya Zulu	28	8			1
St. Monicas'	31	16	1	2	1
Chizongwe	24	14			1
Totals	128	58	13		

2.2 Instruments

In this study, three instruments were designed to collect data as follows:

(i) The Interpretation and Translation of Chemistry Representations Activity (ITCRA), which consisted of five sections, A, B, C, D and E, of which:

- Section A collected information about the experiments (macro level) to which the pupils had been exposed, and how these had been carried out (teacher or pupil demonstrated, pupils did the experiment, or teacher explained what is expected in that experiment).
- Sections B, C and E (part II) assessed how the pupils interpreted macro, symbolic and sub – micro chemistry representations.
- Section D assessed the ability to write a chemical equation (symbolic) from a chemical description (macro).
- Section E (part I) assessed the ability to write a balanced chemical equation from particle diagrams.

(ii) Structured questionnaire for chemistry teachers, which consisted of three sections:

- Section A collected information on the teachers' knowledge of the three chemistry representations, how often each representation was used in teaching, and the strategies used in teaching the different representations.
- Section B (I) collected information about the experiments (macro level) the teachers had exposed their pupils to and how these had been carried out (teacher or pupil demonstrated, pupils did experiment or teacher explained what is expected in that experiment).
- Section B(II) and (III) collected information on strategies used to help pupils visualize atoms and molecules (sub-micro) and understand chemical symbols, formulas and equations, respectively.
- Section C assessed the availability and use of the ball and stick models in schools.

(iii) Focus group discussion (FGD) guidelines. In focus group discussions, pupils were shown ball and stick models and how they can be used to discuss various topics in chemistry. Information was sought on whether they were familiar with these and whether their teachers had used them in any chemistry topic.

The formulated instruments were pilot tested at Chassa Secondary School, a boys' school in the Sinda district of Eastern province of Zambia that had pupils taking chemistry and science chemistry. No change was made to any instrument after the pilot study. With the pilot study carried out at Chassa Secondary School, all the three instruments designed were found to be effective.

Using the ITCRA, data was collected from 128 grade 11 Science 5124 pupils and 58 Chemistry pupils. Using teachers' structured questionnaires, data was collected from a total of 13 teachers of Science 5124 and Chemistry 5070: nine questionnaires were filled in by Science 5124 teachers and nine by Chemistry 5070 teachers. From the 13 teachers, 5 teachers taught both science chemistry and chemistry and filled in two questionnaires each. Focus group discussions were conducted at all five schools using 10 pupils, five from Chemistry 5070 and five from Science Chemistry class.

3 Results and Disussions

The next sections offer a summary of pupils' responses to questions in the ITCRA which assessed pupils' knowledge of interpretation and translation of chemistry representations. The percentage of pupils who gave an incorrect result in an interpretation or translation was computed and the results are presented in bar charts.

3.1 Interpreting Some Macro Chemistry Representations

The Interpretation and Translation of Chemistry Representations Activity (ITCRA) collected information about the experiments (macro level) to which the pupils had been exposed and how these had been carried out (teacher or pupil demonstrated, pupils did experiment, or teacher explained what is expected in that experiment).

There are 21 experiments in both the Science 5124 and Chemistry 5070 syllabi, and six of these were purposely selected and included in section B of the ITCRA. The main selection criterion was that the chemicals and equipment needed to carry out the activities were readily available in the schools.

Pupils were asked to state what would be observed when:

I. An acid is tested with litmus indicator
II. An insoluble salt is made from two soluble salts
III. A carbonate is reacted with an acid
IV. Hydrogen is tested with a lighted match
V. Smoke particles are illuminated in a smoke chamber
VI. The melting point temperature of a pure substance is reached in a cooling curve

The percentage of Science 5124 and Chemistry 5070 pupils who gave an incorrect expected observation was computed and is shown in the bar chart in Fig. 8.1. Table 8.2 shows the mode of instruction that was used to teach the six macro representations above. The percentage of pupils who had done the experiment or seen a demonstration of the experiment was computed for each experiment for Science 5124 and Chemistry 5070 pupils. All the experiments were otherwise taught by lecture method. It is observed from Table 8.2 that teachers in general gave the Chemistry 5070 pupils an opportunity to do or observe an experiment. It is deduced that this is partly due to the fact that teachers know that Chemistry 5070 pupils have

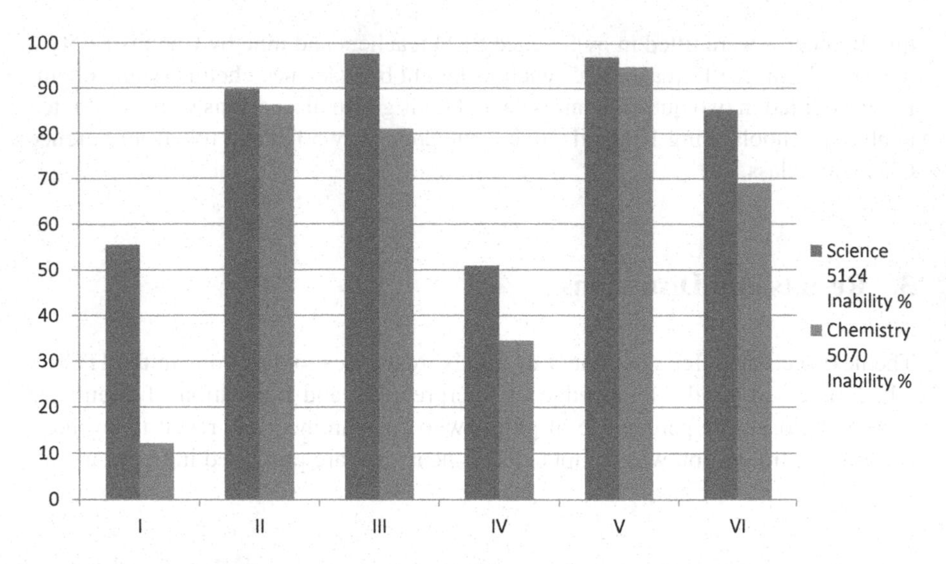

Fig. 8.1 Percentage of pupils who gave an incorrect result in interpreting some macro chemistry representation, (codes for I to VI as described in text)

Table 8.2 Percentage of pupils who had done or seen a demonstration of the selected experiments

Experiment	Science 5124 pupils	Chemistry 5070 pupils
I. Litmus test	11%	100%
II. Precipitation	1%	55%
III. Carbonate-acid	2%	26%
IV. Hydrogen test	20%	40%
V. Smoke chamber	0	1%
VI. Cooling curve	0	3%

to write an external practical examination at the end of grade 12, while the Science 5124 pupils do not.

The results show high inability percentages, with the Chemistry 5070 pupils showing better performance than the science 5124 pupils. An analysis of Table 8.2 shows that the mode of instruction for the six experiments for Science 5124 pupils was almost 100% by lecture mode, i.e. the teacher explained the expected observations. The exception was the hydrogen test experiment, where 20% of the Science 5124 pupils had done the experiment in the laboratory. A comparison of Table 8.2 and Fig. 8.1 suggests that pupils are more likely to recall what they have learned when, in addition to the teacher explaining what is expected, the pupils actually see the outcome.

3.2 Interpreting Some Symbolic Chemistry Representation

Interpretation of symbolic chemistry representations was carried out in section C of ITCRA. Pupils were asked to:

I. Interpret the difference between O and O_2
II. Interpret the difference between H_2O and 2 H_2O
III. Identify chemical symbols and chemical formulas from a given list (Al, CuO, KF, Co, N_2, H, Ne, CO)
IV. Identify elements from a given list (Al, CuO, KF, Co, N_2, H, Ne, CO)
V. Interpret the difference between Co and CO
VI. Calculate the number of atoms in $Al_2(CO_3)_3$ and $(NH_4)_2SO_4$

The percentage of pupils who gave an incorrect result in interpreting some symbolic chemistry representation is given in Fig. 8.2.

In general, Science 5124 pupils could not correctly interpret the differences between O and O_2 or H_2O and $2H_2O$. Inability percentages in both cases were 76%. On the other hand, the Chemistry 5070 pupils performed better, with inability percentages of 59% and 43% for I and II, respectively. Inability percentages in interpreting the difference between Co and CO were generally high, 88% for Science 5124 and 64% for Chemistry 5070. The inability percentage for distinguishing between chemical symbols and chemical formulas and between elements and compounds were greater than 90% in both Science 5124 and Chemistry 5070. Most pupils were not able to count the number of atoms in the given chemical formulas, the inability percentages being 86% for Science 5124 and 76% for Chemistry 5070.

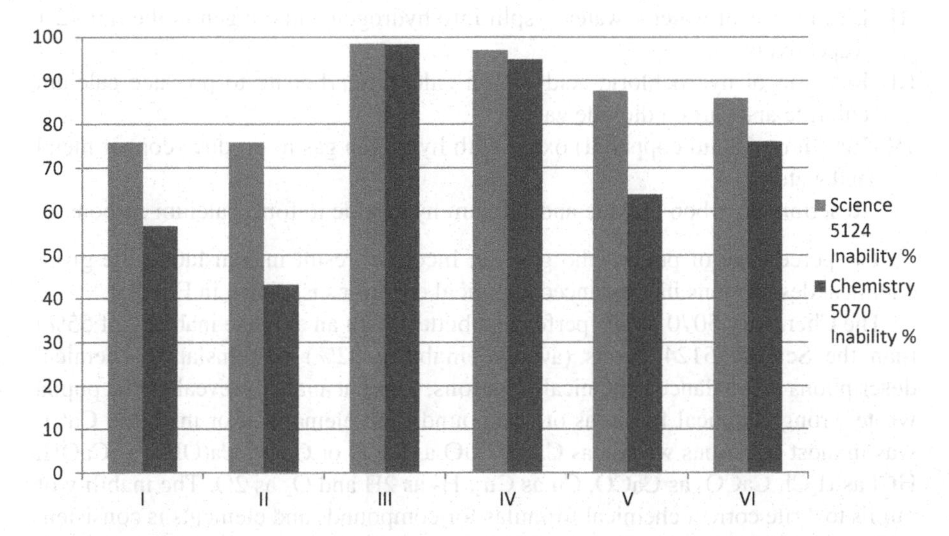

Fig. 8.2 Percentage of pupils who gave an incorrect result in interpreting some symbolic representation, (codes for I to VI as described in text)

Content analysis revealed that a number of pupils wrongly referred to O as *oxygen* or as *one oxygen molecule* and still others as *oxide*. A number of pupils referred to O_2 as *two atoms of oxygen combined* and others as *two molecules of oxygen* and still others as simply *oxygen*. In the case of H_2O and $2H_2O$, a number of pupils referred to H_2O as *water atom*, others as *two atoms of hydrogen and one chemical*, others as *two atoms of hydrogen reacting with oxygen* and still others as simply *water*. A number of pupils referred to $2H_2O$ as *two atoms of water*, others as *four atoms of hydrogen and two of oxygen*, others as *hydrogen and water* and still others as *steam*. As for Co and CO, a number of pupils referred to *Co as carbon and oxygen*, others as *copper* and others as *carbon monoxide* and still others as one *atom of Co*; in the case of CO, pupils referred to it as *carbon dioxide*, others as *cobalt* and others as *one carbon atom and one oxygen atom*. It is apparent from these statements that most pupils had not mastered the basic concepts of atom and molecule of an element on one hand, and the concepts of molecule of an element and molecule of a compound. Some pupils classified N_2 and Co as compounds while KF, CO and CuO were classified as chemical symbols.

3.3 Translation of Some Chemical Descriptions into Balanced Chemical Equations

Translation of chemical descriptions into balanced chemical equations, i.e. symbolic representations, were carried out in ITCRA section D. Pupils were asked to write balanced chemical equations for the following reactions:

I. Reaction of hydrogen with oxygen to produce liquid water
II. Electrolysis of water – water is split into hydrogen and oxygen in the ratio 2:1 respectively
III. Reaction of hydrochloric acid with a calcium carbonate to produce calcium chloride and carbon dioxide gas.
IV. Reaction of solid copper(II) oxide with hydrogen gas to produce copper metal and water
V. Reaction of carbon dioxide and calcium hydroxide to form calcium carbonate.

The percentage of pupils who gave an incorrect result in translating the given chemical descriptions into balanced chemical equations is shown in Fig. 8.3.

The Chemistry 5070 pupils performed better (with an average inability of 55%) than the Science 5124 pupils (average inability 82%) in translating chemical descriptions into balanced chemical equations. Content analysis revealed that pupils wrote wrong chemical formulas of compounds and elements. For instance, $CaCl_2$ was in most equations written as CaCl, CuO as Cu_2O or CuO_2, $Ca(OH)_2$ as CaOH, HCl as H_2Cl, $CaCO_3$ as CaCO, Cu as Cu_2, H_2 as 2H and O_2 as 2O. The inability of pupils to write correct chemical formulas for compounds and elements is consistent with the previously mentioned finding that pupils had not mastered the basic concepts of atoms and molecules, and of elements and compounds.

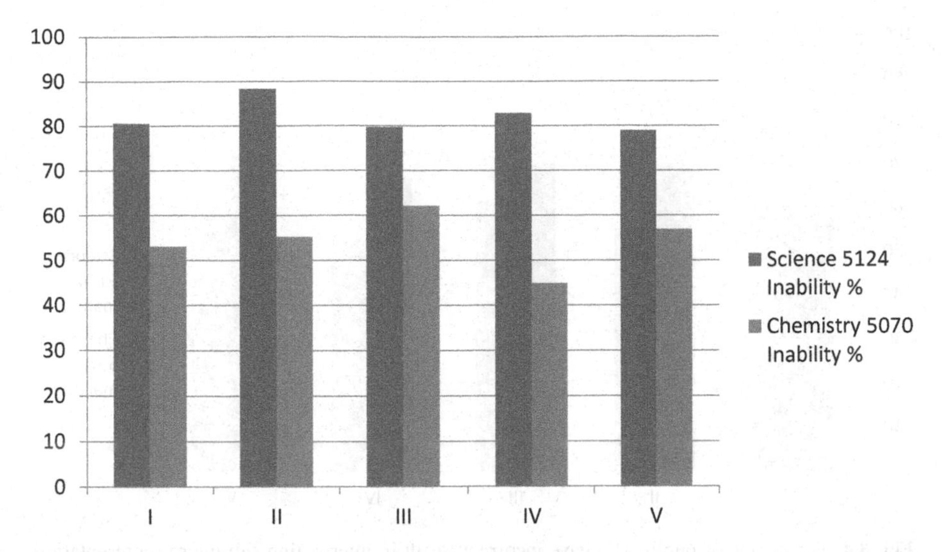

Fig. 8.3 Percentage of pupils who gave incorrect result in translating chemical descriptions into balanced chemical equations (codes for I to V as described in text)

3.4 Interpretation of Some Sub-micro Chemistry Using Particle Diagrams

Interpretation of sub-microscopic chemistry was carried out in ITCRA section E(II) in which pupils were asked to draw particle diagrams of:

I. Five atoms of oxygen and four atoms of hydrogen
II. One molecule of oxygen and one molecule of hydrogen
III. Three molecules of oxygen and two molecules of hydrogen
IV. One molecule of water
V. Three molecules of water.

Figure 8.4 is the bar chart that shows the percentage of pupils who gave an incorrect result in interpreting some sub-micro chemistry representation using particulate diagrams.

The average inability percentages of pupils who could not draw correct particulate diagrams of atoms and molecules is 67% for Science 5124 and 59% for Chemistry 5070. Content analysis of particulate diagrams drawn by pupils of either Science 5124 or Chemistry 5070 revealed that some pupils could draw a molecule of oxygen or hydrogen, but could not draw two or more atoms of oxygen or hydrogen. Other pupils could draw one molecule of oxygen or hydrogen, but not draw two or more molecules of oxygen or hydrogen. In other cases, pupils could draw a molecule of oxygen or hydrogen, but could not draw a molecule of water. Others could draw a molecule of water, but could not draw a molecule of oxygen or hydrogen. Some pupils could draw molecule of water, but could not draw two or more molecules of water. Once again it was noted that pupils are struggling with the concepts of atoms and molecules.

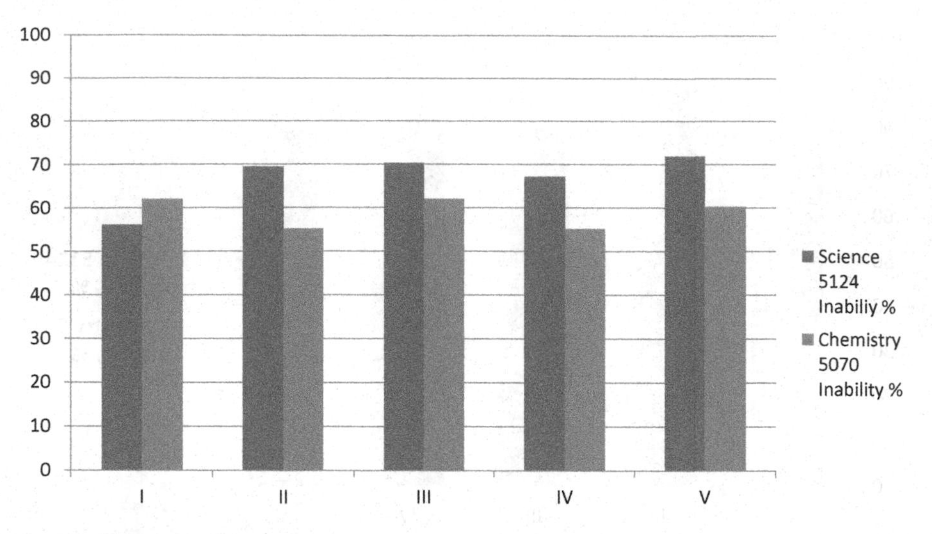

Fig. 8.4 Percentage of pupils who gave incorrect result in interpreting sub-micro representation (codes for I to V as described in text)

From the foregoing, it is apparent that grade 11 pupils in Science 5124 and Chemistry 5070 are generally not able to interpret chemistry representations or translate from one representation to another.

3.5 Translation of an Equation in a Particle Diagram into a Balanced Chemical Equation

Pupils were given the particle diagram (sub-micro) shown in Fig. 8.5 and asked to write the corresponding balanced chemical equation (symbolic).

The correct response was

$$C + O_2 \rightarrow CO_2$$

The percentage of Science 5124 pupils who gave an incorrect result was 89% and that of Chemistry 5070 pupils was 62%. Some of the incorrect answers are given in Box 8.1.

Box 8.1: Some of the Incorrect Answers Given to the Question Shown in Fig. 8.5

$2C + 4O \rightarrow 2C_2O_4$

$C + 2O \rightarrow CO_2$

$C_2 + 4O \rightarrow 2CO_2$

$2C + 4O \rightarrow 2CO_2$

$C_2 + O_4 \rightarrow 2CO_2$

$C + 2O \rightarrow CO_2$

$C_2 + O \rightarrow C_2O$

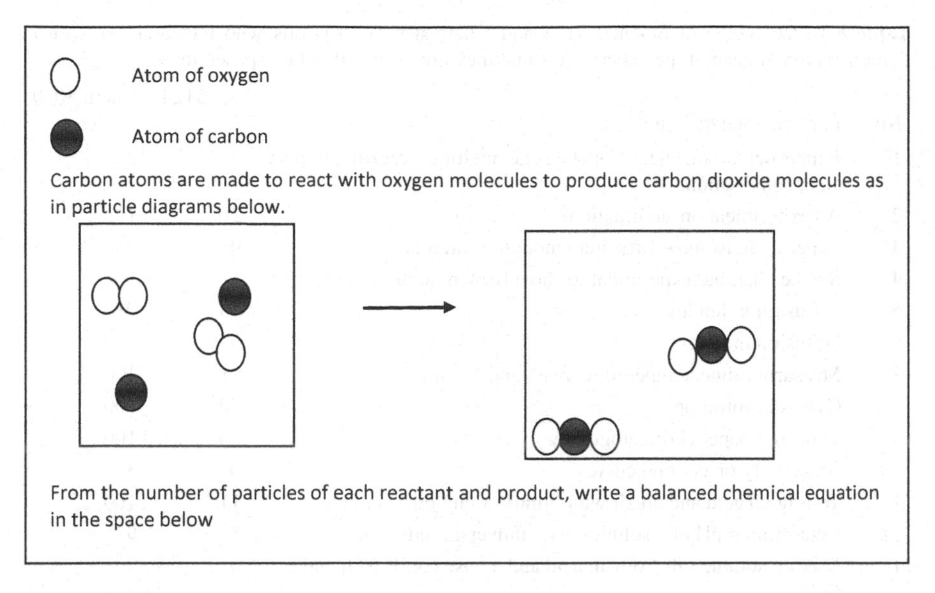

Fig. 8.5 Particle diagrams proposed to students, asking them to translate them into a balanced chemical equation

Once again it is apparent that most pupils had not mastered the basic concepts of atoms and molecules. Furthermore, pupils could not correctly interpret the particle diagrams and translate them to symbolic format.

3.6 Strategies Teachers Use to Teach Each of the Chemistry Representations

To answer the second research question on what strategies teachers used to teach each of the Chemistry representations, data was obtained from both pupils (from ITCRA and focus groups) and teachers (teachers' questionnaire). It was apparent from ITCRA that the macro aspects were taught mostly by lecture method. Table 8.3 lists the 21 experimental activities that Science 5124 and Chemistry 5070 pupils are expected to have done by the end of grade 11, and shows the percentage of pupils who claim to have done or seen a demonstration of the experiment. Science 5124 pupils were not exposed to most experimental observations while Chemistry 5070 pupils were exposed to some experiments especially those known to appear in the external practical examination that Chemistry 5070 pupils take at the end of grade 12. Almost all chemistry 5070 had done or seen a demonstration of an acid/base titration and identification of cations and anions, which are standard experiments in grade 12 external examinations. The science 5124 pupils had not seen these experiments. Both pupils and teachers indicated that they do not go on field trips to chemical industries or the environment to study related chemical phenomena.

Table 8.3 Percentage of Science 5124 and Chemistry 5070 pupils who had done or seen a demonstration of each of the experiments that they are expected to be exposed to

S/ No.	Experimental activities	Sci 5124 %	Chem 5070 %
1	Experiment(s) on change of states i.e. melting, freezing, boiling and condensation.	8	22
2	An experiment on sublimation.	2	19
3	Experiment to show Brownian motion in liquids	0	0
4	Smoke chamber experiment to show Brownian motion in gases.	1	1
5	Diffusion in liquids	33	91
6	Diffusion in gases	9	0
7	Measuring time, temperature, mass and volume.	91	100
8	Carry out filtration	70	100
9	Carry out paper chromatography	70	100
10	An activity on cooling curves	0	3
11	Testing an acid and alkali with litmus or any other indicator	11	100
12	Determining pH of a solution with universal indicator	5	9
13	Making soluble salt from an acid and a base e.g. H_2SO_4 and CuO	0	7
14	Making soluble salt from an acid and alkali e.g. H_2SO_4 and NaOH (Titration)	2	100
15	Making an insoluble salt from two soluble salts (precipitation)	1	55
16	Soluble salt from an acid and metal and from a carbonate	2	26
17	Identifying metallic ions	2	100
18	Detecting nitrates, sulphates, chlorides and carbonates in solutions.	2	100
19	Preparing and testing carbon dioxide	24	43
20	Preparing and testing oxygen	23	43
21	Preparing and testing hydrogen	20	40

From these findings, it would seem that experiments are only made available to pupils to enable them to pass their grade 12 examination. Performing laboratory experiments and or field observations is not a standard method of teaching chemistry. Teachers miss out on the opportunity to make pupils active learners and interest them in the subject. That the macro representations are routinely taught by lecture method is at odds with the basic understanding of a macro representation, that it is an experiential level, whereby a pupil makes an observation either in the laboratory or in the field (Johnstone 2000). Hence, one can ask the question whether or not this can be taught by lecture. The poor performance of pupils shown in Fig. 8.1 may be a reflection of the fact that they had no experience with experimental activities.

Both the teachers (in the teacher's questionnaire) and pupils (in ITCRA) indicated that symbolic chemistry was taught mainly by lecture. From ITCRA, pupils indicated that sub-micro chemistry was taught through use of structural formulas, cross dot diagrams, ionic structures and particulate diagrams. In isolated cases, some pupils indicated some use of computers in learning sub-microscopic chemistry. There was no indication from the pupils of either Science 5124 or Chemistry 5070 that ball and stick models or work cards or games or songs and poetry or role play and drama have ever been used in their learning of chemistry.

3.7 *How Often Each of the Three Chemistry Representations Were Used in the Teaching and Learning of Science 5124 and Chemistry 5070*

All the teachers in the study indicated that they were aware of the three representations of chemistry. Teachers were asked how often in the teaching and learning of Science 5124 and Chemistry 5070 the following activities carried out:

- Experimental activities from which pupils make observation (macro representation).
- Activities that enabled pupils to visualize atoms, molecules and ions (sub-micro representation).
- Activities that enabled pupils to master chemical symbols and formulas of elements and compounds and chemical equations of various reactions (symbolic representation).

Table 8.4 shows how often each chemistry representation was used by the teachers to teach Science 5124 and Chemistry 5070.

Table 8.4 shows that almost all the teachers in the study indicate that the symbolic representation activities were carried out most frequently. All the teachers of Chemistry 5070 claim to have carried out sub-micro activities often or very often, compared two thirds of Science 5124 teachers. Noteworthy is that 44% of the Science 5124 teachers indicated that they rarely carried out experimental activities (macro level). This is consistent with the previous observations (Table 8.3), where it was noted that the Science 5124 pupils were hardly exposed to experimental observations.

Table 8.4 Percentage of teachers indicating how often they used each chemistry representation in the teaching of chemistry. (Science 5124 teachers, N = 9 and Chemistry 5070 teachers, N = 9)

	Percentage of teachers indicating how often they used each of the chemistry representations									
	Very often		Often		Rarely		Very rarely		Never	
Chemistry representation	Scie 5124	Chem 5070	Scie 5124	Chem 5070	Scie 5124	Chem 5070	Scie 5124	Chem 5070	Scie 5124	Chem 5070
Macro	11.1	14.3	44.4	57.1	44.4	14.3	0	14.3	0	0
Symbolic	88.9	100	11.1	0	0	0	0	0	0	0
Sub-micro	55.6	85.7	11.1	14.3	33.3	0	0	0	0	0

3.8 Determining the Extent to Which the Ball and Stick Models Are Used in the Teaching and Learning of Science 5124 and Chemistry 5070

Data on the extent of use of ball and stick models was collected from both teachers (the teachers' questionnaire) and pupils (ITCRA and focus groups). From the teachers' questionnaire, 10 out of 13 teachers indicated that the ball and stick models were available in their schools and that they had used them to teach sub-micro chemistry in organic chemistry topics, covalent bonding and making chemical formulas. From the pupils' questionnaire in the ITCRA, almost all pupils indicated that their teachers never took ball and models to class, at least up to the time of the study which was at the end of grade 11. From the Ball and Stick Chemistry Visualisation Activity (BaSCVA) in focus group discussions, all pupils in the five schools indicated that, up to the time of the study, no chemistry teacher had used the ball and stick models in teaching any chemistry topic. However, some pupils had witnessed the use of the ball and stick models on some education programmes on television.

Based on these findings, it is clear that teachers of both Science 5124 and Chemistry 5070 had not used the ball and stick models to teach sub-microscopic chemistry, at least up to the time when this study was conducted. In the teachers' questionnaire, all teachers indicated that the ball and stick models are generally meant to be used in teaching topics in organic chemistry, which is normally taught in grade 12. The study findings suggest that teachers are not conversant with the fact that ball and stick models can be used to teach many topics in chemistry. The teachers in the study did not make an effort to help pupils develop mental pictures of atoms and molecules, and this may be contributing to pupils' lack understanding of the basic concepts of atom of an element and molecule of element or compound.

4 Conclusions

This study has established that grade 11 chemistry pupils could not adequately express themselves at symbolic, sub-microscopic and macroscopic levels. At both symbolic and sub-microscopic levels, pupils had not grasped the concepts of atom(s) of element and molecules of elements or compounds. At macroscopic level, most pupils were not able to state expected observations in basic experiments. The study found that, in general, the macroscopic representation was taught by the lecture method. It is a finding of this study that Science 5124 pupils were generally not confronted with the macro representation (experiments) and that Chemistry 5070 pupils were normally shown only those macro representations (experiments) that teachers expected to appear in the external grade 12 examinations. It is concluded that the macro representation is not adequately taught.

The study also found that pupils could not correctly interpret chemical symbols and chemical formulas. Pupils could not distinguish between atoms and molecules of an element or between molecules of elements and compounds, and had not mastered the concept of diatomic molecules. This was despite the fact that the bulk of teaching occurred at the symbolic level. This underlies the importance of teaching chemistry at all three levels.

In order to ensure that effective learning takes place in schools, it is recommended that teachers be equipped through both pre-service and in-service training, workshops and seminars to teach chemistry at all three levels.

References

Bradley, J. D. (2014). The chemist's triangle and a general systematic approach to teaching, learning and research in chemistry education. *African Journal of Chemical Education, 4*(2), 64–79.

Bruner, J. S. (1977). *The process of education*. Cambridge: Harvard University Press.

Curriculum Development Centre. (2013a). *Chemistry 5070 syllabus: Grades 10–12*. Lusaka: Author.

Curriculum Development Centre. (2013b). *Science 5124 syllabus: Grades 10–12*. Lusaka: Author.

Davidowitz, B., Chittleborough, G., & Murray, E. (2010). Student-generated submicro diagrams: A useful tool for teaching and learning chemical equations and stoichiometry. *Chemistry Education Research and Practice, 11*, 154–164.

Gabel, D. (1999). Improving teaching and learning through chemistry education research: A look to the future. *Journal of Chemical Education, 76*(4), 548–554.

Gilbert, J. K., & Treagust, D. (2009). Introduction: Macro, sub-micro and symbolic representations and the relationship between them: Key models in chemical education. In J. K. Gilbert & D. Treagust (Eds.), *Multiple representations in chemical education* (pp. 1–8). Dordrecht: Springer.

Johnstone, A. H. (1991). Why is science so difficult to learn? Things are seldom what they seem. *Journal of Computer Assisted Learning, 7*, 75–83.

Johnstone, A. H. (2000). Teaching of chemistry – Logical or psychological? *Chemistry Education: Research and Practice in Europe, 1*(1), 9–15.

Johnstone, A. H. (2006). Chemical education research in Glasgow in perspective. *Chemistry Education Research and Practice, 7*(2), 49–63.

Joseph, A. (2011). *Grade 12 learners' conceptual understanding of chemical representations*. Retrieved June 8, 2015 from https://ujdigispace.uj.ac.za/bitstream/handle/10210/6575/Joseph.pdf?sequence=1

Kozma, R. B., & Russell, J. (1997). Multimedia and understanding: Expert and novice responses to different representations of chemical phenomena. *Journal of Research Science Teaching, 34*(9), 949–968.

Nyachwaya, J. M., & Wood, N. B. (2014). Evaluation of chemical representations in physical chemistry text books. *Journal of Chemistry Education and Practice, 15*, 720–728.

Taber, K. S. (2013). Revisiting the chemistry triplet: Drawing upon the nature of chemical knowledge and the psychology of learning to inform chemistry education. *Chemistry Education Research and Practice, 14*(2), 156–168.

Talanquer, V. (2011). Macro, submicro, and symbolic: The many faces of the chemistry "triplet". *International Journal of Science Education, 33*(2), 179–195.

Treagust, D. F., Chittleborough, G., & Mamiala, T. L. (2003). The role of the submicroscopic and symbolic representations in chemical explanations. *International Journal of Science Education, 25*(11), 1353–1368.

Wu, H. K., Krajcik, J. S., & Soloway, E. (2001). Promoting understanding of chemistry representations: Students' use of a visualization tool in the classroom. *Journal of Research and Science Teaching, 38*(7), 821–842.

Chapter 9
The Project IRRESISTIBLE: Introducing Cutting Edge Science into the Secondary School Classroom

Jan Apotheker

1 Introduction

The project IRRESISTIBLE is the result of a proposal that was accepted within the FP-7 program of the EU under number 612367. The proposal was made in response to call SIS.2013.2.2.1.: (Anonymous 2012)

Area 5.2.2.1. Supporting formal and informal science education in schools as well as through science centres and museums and other relevant means.
SiS.2013.2.2.1-1: Raising youth awareness to Responsible Research and Innovation through Inquiry Based Science Education.

Within the project IRRESISTIBLE, activities are designed that foster the involvement of students and the public in the process of responsible research and innovation. The project raises awareness about RRI in two ways:

- Increasing content knowledge about research by bringing topics of cutting-edge research into the program
- Fostering a discussion among the students regarding RRI issues about the topics that are introduced.

Both formal and informal learning environments play an important role in these activities. The project functions within the field of chemistry education and tries to bridge the gap between research and secondary schools. By having the students make an exhibition about their projects which are exhibited in science centers, the project also involves the general public.

A number of new aspects came together in the project. In the methodology used for the project, the first aspect is the teacher professionalization, the second is the

J. Apotheker (✉)
Faculty of Science and Engineering, University of Groningen, Groningen, Groningen, The Netherlands
e-mail: J.H.APOTHEKER@RUG.NL

L. Mammino, J. Apotheker (eds.), *Research in Chemistry Education*,
https://doi.org/10.1007/978-3-030-59882-2_9

use of contexts and the 5E methodology, the third is the use of exhibits as a means of assessing the learning process of the students as well as communicating the results of the project. One last aspect is the combination of formal and informal learning in the design of the project. The results of the project will be disseminated in various articles. This chapter discusses both the teacher professionalization and the use of contexts and the (adapted) 5E-methodology.

2 The Content of the Educational Material

The content that is used in the educational material is new. Topics in science research that are currently subject of research at the universities involved in the project have been chosen as subject for the educational material. Table 9.1 gives an overview of the chosen topics and Fig. 9.1 shows the front page of the module from the University of Palermo.

3 Responsible Research an Innovation

As part of the content material, the idea of Responsible Research and Innovation (RRI) is introduced. RRI has been a focal point within the EU for the last 10 years, in the Framework programs dealing with Science in Society, as well as in the Horizon 2020 programs. The main idea behind RRI is discussed by Sutcliffe

Table 9.1 Titles and subjects of the educational material produced

Country	Title of educational material	Science subject
Portugal	Geo engineering and climate control	Geo-engineering
	Evaluation of Earth health through polar regions	Eco systems
Finland	Atmosphere and climate change	Grätzel cells
Turkey	Nano and health	Application of silver nano particles as bacteriostat
Poland	The catalytic properties of nanomaterials	Role of nanomaterials as catalyst
The Netherlands	Carbohydrates in wind-energy	Specific carbohydrates
Italy, Bologna	Nanotechnology for solar energy	Photo voltaic cells
	Nanotechnology for information by exploiting light/matter interaction	Luminescent nano sensors
Italy, Palermo	Energy sources	Nano technology
Israel	The RRI of Perovskite based photovoltaic cells	Perovskite cells
Germany	Oceanography and climate change	Off shore wind energy
	Plastic, bane of the oceans	Plastic waste in the oceans
Greece	Nanoscience applications	Nanotechnology

Fig. 9.1 Cover page of the module from Palermo (Italy)

(Sutcliffe 2011), and focuses on the relationships between society and research; it can be summarised as follows

- The deliberate focus of research and the products of innovation to achieve a social or environmental benefit.
- The consistent, ongoing involvement of society, from beginning to end of the innovation process,
- Involvement of the public & non-governmental groups, who themselves are mindful of the public benefit.

This has been reset to six key issues that have been chosen in the project as a guideline for the introduction of RRI in the classroom (Anonymous Leaflet RRI 2015), namely:

- Engagement
- Gender equality
- Science education
- Ethics
- Open Access
- Governance

Each of these RRI issues is linked to the science subject discussed. The key issues are explained in Table 9.2.

Table 9.2 The key issues of RRI (taken from Anonymous)

1. Engagement
The first key to RRI is the engagement of all societal actors – researchers, industry, policymakers and civil society – and their joint participation in the research and innovation process, in accordance with the value of inclusiveness, as reflected in the Charter of Fundamental Rights of the European Union. A sound framework for excellence in research and innovation entails that the societal challenges are framed on the basis of widely representative social, economic and ethical concerns and common principles. Moreover, mutual learning and agreed practices are needed to develop joint solutions to societal problems and opportunities, and to pre-empt possible public value failures of future innovation.
2. Gender Equality
Engagement means that all actors – women and men – are on board. The under-representation of women must be addressed. Research institutions in particular their human resources management, need to be modernized. The gender dimension must be integrated in research and innovation content.
3. Science Education
Europe must not only increase its number of researchers, it also needs to enhance the current education process to better equip future researchers and other societal actors with the necessary knowledge and tools to fully participate and take responsibility in the research and innovation process. There is an urgent need to boost the interest of children and youth in mathematics, science and technology, so they can become the researchers of tomorrow, and contribute to a science-literate society. Creative thinking calls for science education as a means to make change happen.
4. Open Access
In order to be responsible, research and innovation must be both transparent and accessible. This means giving free online access to the results of publicly-funded research (publications and data). This will boost innovation and further increase the use of scientific results by all societal actors.
5. Ethics
European society is based on shared values. In order to adequately respond to societal challenges, research and innovation must respect fundamental rights and the highest ethical standards. Beyond the mandatory legal aspects, this aims to ensure increased societal relevance and acceptability of research and innovation outcomes. Ethics should not be perceived as a constraint to research and innovation, but rather as a way of ensuring high quality results.
6. Governance
Policymakers also have a responsibility to prevent harmful or unethical developments in research and innovation. Through this key we will develop harmonious models for Responsible Research and Innovation that integrate public engagement, gender equality, science education, open access and ethics.

The produced educational material will be made available through the website of the project, http://www.irresistible-project.eu/index.php/en/, as well as www.scientix.eu.

4 Teacher Professionalization

For the design of the educational material, so-called Communities of Learner were formed. A community of learners (CoL), or a community of practice, consists of people involved in the design. In the project, the core group consisted of teachers from secondary schools, with a chemistry education researcher. Other members were representatives from the research groups as well as from the science centre. Loucks-Horsley et al. (2010) indicated that CoLs are a powerful tool for training teachers. Eilks and Markic (2011) also used similar groups. Similar CoLs have been used for teacher training both in the 'project nieuwescheikunde' in the Netherlands (Apotheker Invalid date) and in 'Chemie im Kontext' (Nentwig et al. 2007). Figure 9.2 offers an overview of the composition of a typical CoL used in this project?

In the first round of the project, the work in the CoL is focused on the design of the educational material. In the new material, formal and informal learning environments are connected.

Informal environments can be used in different ways in the educational process:

- Attract positive attention from students towards a subject,
- Introduce content knowledge in a different way,
- Possibility to discuss with stakeholders about RRI issues.

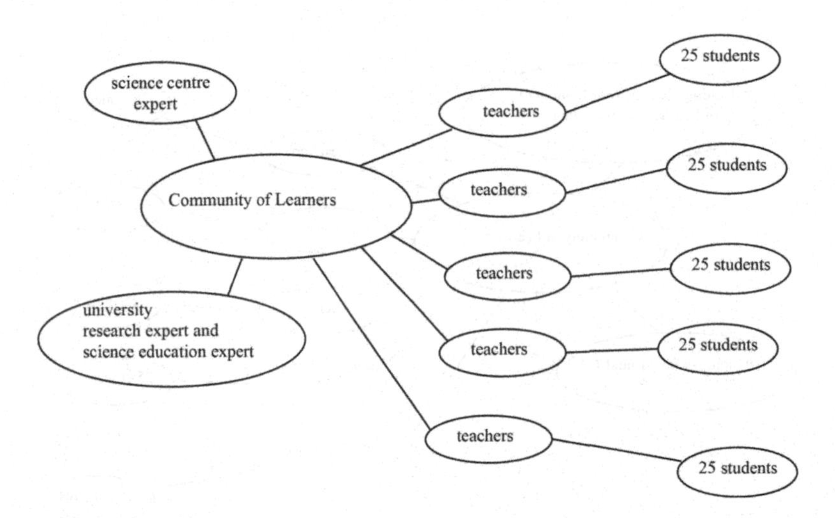

Fig. 9.2 Composition of the community of learners in the first round of the project presented in this chapter

In the formal learning environment, the teachers are expected to adapt existing material into a new format in which

- students are activated,
- interest from both boys and girls is promoted,
- students take responsibility for their own learning,
- new topics are introduced that demonstrate the overlap between different fields of science.

In the second round the role of the teachers changes from that of a developer into that of a coach (Fig. 9.3).

Each teacher now coaches 5 other teachers in the use of the developed material. In this second round, the module developed by the teachers is used, as well as at least two modules developed by another partner. The experience gained in this second round is used to improve and edit the material, up to the point that teachers only need to make small changes to use the material directly in their classroom.

5 Work of the CoLs in the First Round

The work in the first round of the CoL's differed from country to country. The procedure used in the Netherlands is described in the following paragraphs and subsections, to serve as an illustrative example.

The CoL comprised 9 biology and chemistry teachers from secondary schools and was led by an expert in chemistry education research and an expert from Science LinX, the outreach organization of the University of Groningen (www.rug.nl/sciencelinx). The teachers volunteered to take part in it. Science LinX negotiated a

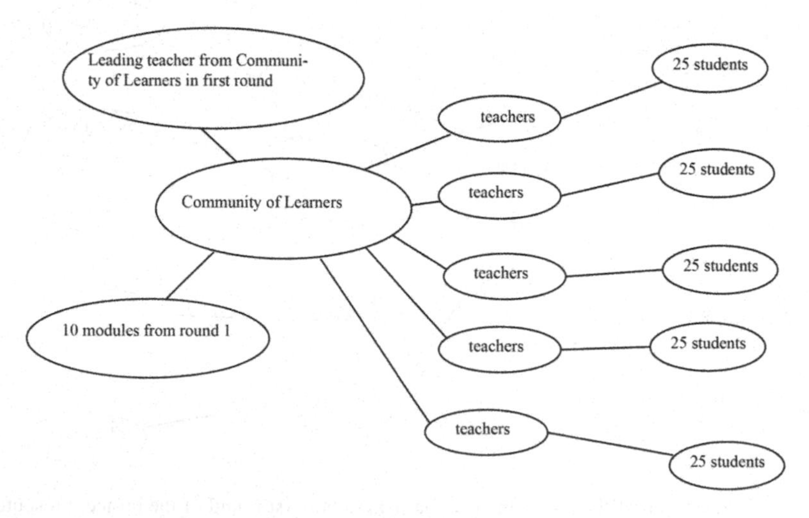

Fig. 9.3 Community of learners in the second round

contract with the schools, in which the teachers were given 40–80 h to work on the project, in return for the schooling the teachers would receive. The meetings were organized from 16:00 till 20:00 in a central town in the Netherlands, so that teachers could travel easily.

The meetings were divided in three groups:

- Introductions (5 meetings)
  - Introduction to the science content
  - Introduction about the use in industry
  - Introduction on RRI
  - Introduction on the use of the 6 E method
  - Introduction on the use of exhibits
- Design and writing of the material (about 10 meetings)
  - Decision on the content
  - Chapters to be included
  - Relationship with curriculum
  - Regular feedback on material
- Try out of the material (about 5 meetings)
  - Feedback and rewriting of the module

In general, the schedule of the meetings started with about two and a half hours of work (until 18:30) followed by a dinner break till about 19:00. The last hour was used for taking decisions and dividing the work for the next meeting. Minutes were taken by one of the coordinators.

5.1 Introductions

During the first 5 meetings about 2 h lectures were given on aspects mentioned in the previous list. In the second part of the meeting, the lectures were discussed; the questions arising in the discussion were formulated to be answered in the next session.

In the first session, Lubbert Dijkhuizen (http://www.cccresearch.nl/media/brochures/SNN_Rapportage_Wire_O.pdf) discussed the importance of a specific type of Human Milk Oligosaccharides (HMOS) that play an important role in the development of gut bacteria. The group of professor Dijkhuizen identified these HMOS.

In the second lecture, someone from Friesland-Campina discussed Vivinal®GOS (https://www.vivinalgos.com/en/), a component consisting mainly of Galacto-Oligosaccharides (GOS). This component is added to formula milk that is produced by Friesland-Campina, and has similar properties as HMOS. Friesland-Campina built a plant to produce these GOS (Fig. 9.4) following the research done in this project.

Fig. 9.4 Plant of FrieslandCampina for the production of Galacto-Oligosaccharides. (Taken from https://www.vivinalgos.com/en/news/)

During the lecture, the technology for the production of formula milk, as well as the production of GOS, was introduced. At a later time, the teachers were given the opportunity to visit the plant.

The author gave the other lectures. The lecture about the exhibitions was given at a later stage, when it was needed in the process.

All the content was new to the teachers. They had to familiarize themselves with this new subject, of which they had only some basic knowledge.

5.2 Reflections on the Roles of the Teachers in the Project

The teachers were able to follow the lectures and integrate and use the new knowledge in the design of the module. During the writing process, the gaps in their knowledge of the new material became apparent. These were remedied by easy contact with both the research group and Friesland-Campina.

5.3 The Design Process

The 5E method, as described by Bybee et al. (2006), was used as a basic template for the design of the module.

The 5E method has different steps. In the first three (Engage, Explore, Explain), the content knowledge is studied and learnt. In the last two steps (Elaborate, Evaluate), the focus is on discussing the RRI issues regarding the topic studied. (see Table 9.3).

An extra step (Exchange) was introduced between these last two steps, and involved the development of an exhibition by the students. Students devising and presenting an exhibition is a means o transforming science from product to process (Hawkey 2001). During these exhibits' preparation, learners will ask questions, use logic and evidence in formulating and revising scientific explanations, recognizing and analyzing alternative explanations, and communicate scientific arguments.

Table 9.3 The extended 5E model, incorporating inquiry based science education

Phase	Description	Techniques used
Engage	Getting interested in the subject of the module. Acquire a sense of ownership with the subject. Students want to learn more	Students gather information, are asked questions about breast feeding, find info about the composition of formula milk and breast milk
Explore	In the explore phase of this module pose central questions about what they want to learn in this module	The research about HMOS is introduced. In order to understand the research students need to know about a number of things
Explain.	In the explanation phase knowledge is gained, data collected and scaffolded. The teachers and the students will scaffold the content knowledge	In the biology chapter the role of bacteria in the digestion is discussed. One of the questions is how does a person acquire about 2 kg bacteria? What role do these bacteria play in your health?
		In the chemistry chapter the biochemistry of proteins, fats and carbohydrates is introduced, including HMOS and GOS
		In the engineering chapter the process of the industrial production of GOS is introduced
Elaborate	In the elaboration phase the attention shifts to RRI questions. Students will confront researchers with challenges to be answered by the scientists	In this module advertising for formula milk is discussed as an example of RRI issues.
		Scandals about formula milk (Nestlé, China) are discussed. A debate is organized by the students
Exchange	One of the assignments will be the design of an exhibit, which will be displayed in the science Centre. Posters or other presentation modes may also be used.	Exhibits using an IKEA cupboard have been made and displayed during the night of arts and sciences.
Evaluate	In the evaluation phase the students are tested on their content knowledge.	Students take a written test for a summative assessment
	The module is evaluated as well	

Through the construction and presentation of exhibits on Responsible Research and Innovation, both teachers and students are introduced to a different type of science from the one that is usually presented in science classes. Most of the formal science education focuses on a conventional, non-controversial, established and reliable science.

In the design of the module, the model of educational reconstruction (Duit et al. 2012) was used. The teachers discussed which parts of the secondary school curriculum fitted best with the science content. This led to the topics mentioned in Table 9.3. The way these topics are introduced extends beyond the regular curriculum, because the new science needed to be introduced as well.

The chapter about exhibits was completely new for the teachers. Several teachers attended a workshop about the exhibits, given at the University of Lisbon, one of the partners in the IRRESISTIBLE project. This gave enough insight to develop an exhibition guide for teachers. This guide is downloadable from the website of the project.

During the project, case studies were done by the local groups, according to a format designed by the University of Lisbon. The information gathered in these case studies indicated that, even though both students and teachers were hesitant about making the exhibitions, they were very pleased with the final results. They were uncertain about the new science they learnt, and about the concepts of Responsible Research and Innovation they were meant to introduce.

In a later stage, the results of these case studies will be reported in deliverables 3.3 and 3.4. These can be found on the projects website.

In Figs. 9.5, 9.6 and 9.7, photographs of exhibits are given, showing different types of exhibits.

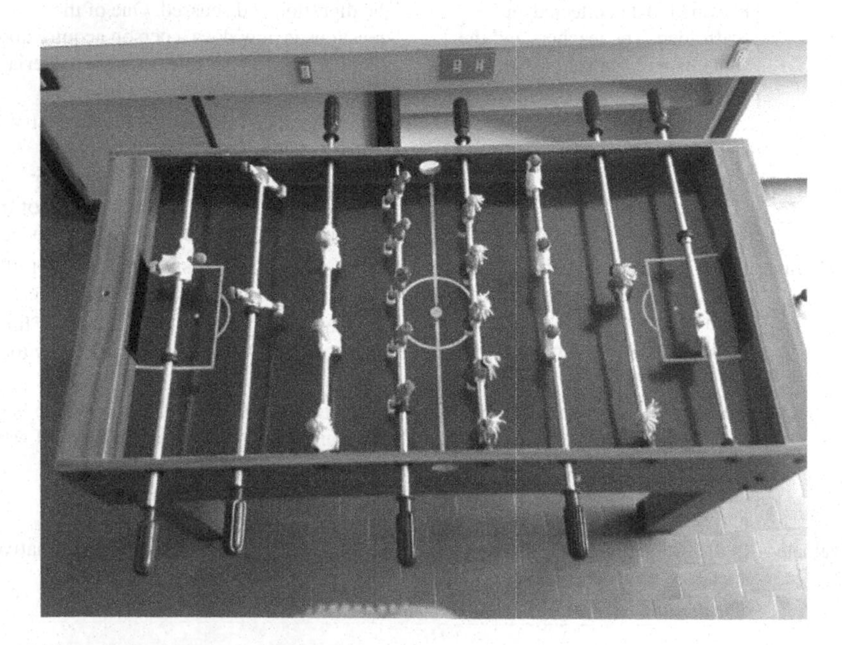

Fig. 9.5 Football game designed by students from Bologna (Italy)

Fig. 9.6 Cartoons designed by students from Lisbon (Portugal)

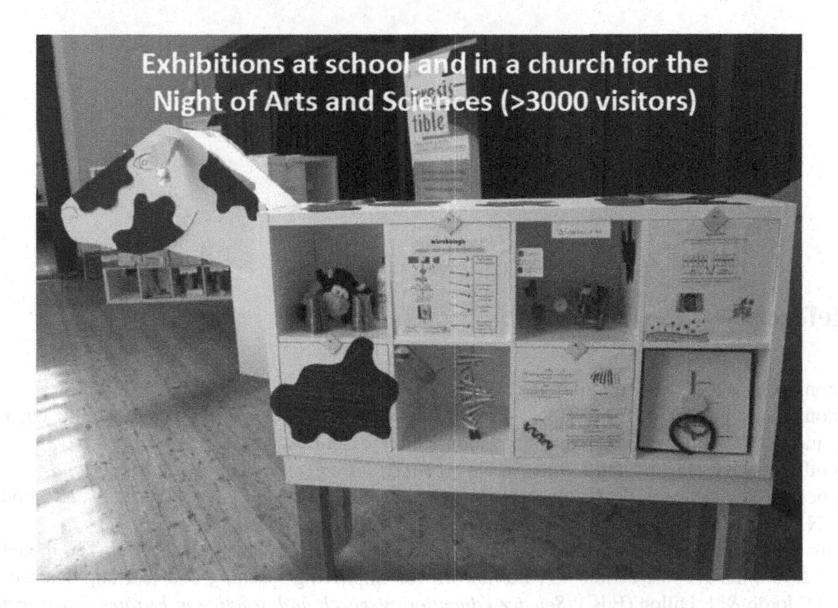

Fig. 9.7 Ikea Exponeer type cupboard designed by students from Groningen (The Netherlands)

6 Conclusions

In the IRRESISTIBLE project, teachers have been able to introduce new science content into the classroom. In the case of The Netherlands, this included an innovation that led to the improvement of formula milk. The way this innovation was realized in industry was part of the project.

Even though RRI is a relatively new concept, teachers have been able to introduce it into their modules. This can be seen most specifically in the design of the exhibits, which have proven to be a powerful learning tool for students, mainly because they must decide how they are going to present the new science they have learnt to the public. In addition, they have to link the science to RRI concepts.

Lastly, the exhibitions bring out the creativity of students in a very nice way. They demonstrate that they are able to come up with some very nice intricate designs.

The project demonstrated that current research can be introduced at secondary school level. It gives students a better idea of the role of chemistry, and of chemists, in the development and innovation in society.

Acknowledgments This project has received funding from the European Union's Seventh Framework Programme for research, technological development and demonstration under grant agreement no 612367.

It is a coordination and support action under FP7-SCIENCE-IN-SOCIETY-2013-1, ACTIVITY 5.2.2 Young people and science: Topic SiS.2013.2.2.1-1 Raising youth awareness to Responsible Research and Innovation through Inquiry Based Science Education.

http://www.irresistible-project.eu/index.php/en/

References

Anonymous. (2012). *Information and notices*, 55: 6.

Anonymous Leaflet RRI. (2015). https://ec.europa.eu/research/swafs/pdf/pub_public_engagement/responsible-research-and-innovation-leaflet_en.pdf. Accessed 6 Oct 2015.

Apotheker, J. H. (Invalid date). *Introducing a context based curriculum in the Netherlands*.

Bybee, R. W., Taylor, J. A., Gardner, A., Van Scotter, P., Powell, J. C., Westbrook, A., & Lamdes, N. (2006). *The BSCS 5E instructional model: Origens and effectiveness*.

Duit, R., Gropengießer, H., Kattmann, U., Komorek, M., & Parchmann, I. (2012). The model of educational reconstruction – A framework for improving teaching and learning science. In D. Jorde & J. Dillon (Eds.), *Science education research and practice in Europe, retrospective and prospective* (pp. 13–39). Rotterdam: Sense Publishers.

Eilks, I., & Markic, S. (2011). Effects of a long-term participatory action research project on science teachers'. *Professional Development, 7*, 149–160.

Hawkey, R. (2001). Innovation, inspiration, interpretation: Museums, science and learning. *Ways of Knowing Journal, 1*, 31–36.

Loucks-Horsley, S., Stiles, K. E., Mundry, S., Love, N., & Hewson, P. W. (2010). *Designing professional development for teachers of science and mathematics* (3rd ed.). Thousand Oaks: Corwin Press.

Nentwig, P. M., Demuth, R., Parchmann, I., Gräsel, C., & Ralle, B. (2007). Chemie im Kontext: Situated learning in relevant contexts while systematically developing basic chemical concepts. *Journal of Chemical Education, 84*, 1439–1444.

Sutcliffe, H. (2011). *A report on responsible research and innovation.*

Chapter 10
Chemistry Teaching and Chemical Education Research: 30-Year Experience in Integration Pathways

Liliana Mammino

1 Introduction

Teaching inherently involves research whenever a teacher wishes to tune teaching approaches to the students' needs and to continuously improve the quality of the approaches (the term *teacher* is here used comprehensively, without distinction of instruction levels). The practical pursuing of these objectives responds to the "action research" paradigms, which Levin (1947, reported in Girod n.d.) outlines as a three-step recursion (or iterative) process, comprising the following steps: planning, which involves reconnaissance; taking actions; and fact-finding about the results of the action.

Action research may have a number of advantages for the teaching activity. The greatest advantage is likely its 'experimental' character. It does not need any *a-priori* framework about teaching and learning (thus avoiding the risk of fitting observations into some pre-existing framework). The investigation component is central in the diagnoses-design-implementations-new diagnoses loop. In other words, direct observation and interpretation are the dominant components. The mode of proceeding is typical of experimental research. The researcher-teacher's mind is completely open while investigating the contextual reality, i.e., identifying as many features as possible about students' difficulties and responses, analysing them for what they communicate about students' actual situation and making inferences about what could be done to improve the quality and effectiveness of learning. In this way, the design of pedagogical options and approaches is closely linked to observations. Familiarity with literature and exchanges with other researchers provide crucial enrichment by helping refine diagnosing, interpretation and design abilities, and comparisons with analogous – or, at least, partially similar – problems and interventions in other contexts can be particularly fruitful.

L. Mammino (✉)
School of Mathematical and Natural Sciences, University of Venda, Thohoyandou, South Africa
e-mail: sasdestria@yahoo.com

L. Mammino, J. Apotheker (eds.), *Research in Chemistry Education*,
https://doi.org/10.1007/978-3-030-59882-2_10

After a brief analysis of the «research + implementation» recursion loop for the integration of teaching and research, the current work presents an overview of the author's direct experience in this regard, through over 30 years of chemistry teaching (the last 20 of which at the University of Venda (UNIVEN) in South Africa); the taught courses comprised first year general chemistry courses, all the physical chemistry courses (for both undergraduates and postgraduates) and the process technology undergraduate course. Specific attention is given to the modes through which diagnoses constitute the basis for the design of interventions, with the features of the design depending on the observations, on the nature of the identified problems and on the teacher's hypotheses on how they can best be tackled.

2 The «Research + Implementation» Loop

The recursion nature of the «research + implementation» loop inherent in the integration of teaching and research can be illustrated in the typical form of a flow chart, with the steps outlined in Fig. 10.1. The first two steps are sort-of initiation steps; the subsequent steps are the ones constituting the loop, as also highlighted by the external arrows on the left of the diagram. Differently from mathematical optimization procedures, this loop is open-ended, as it will never reach what in mathematics is called "convergence" (a result that is satisfactory with respect to a pre-set threshold and, therefore, does not require further iterations). Within teaching and learning, one does not reach an optimal 'final' situation, but aims at continuous improvement (what is 'natural', because the activity deals with human beings, not with mathematical entities). Since teaching and learning pertain to human beings, the 'content' of the recursion steps is not always the same in subsequent loops, because new factors may and do appear with the changing characteristics of new groups of learners in the same context and, therefore, subsequent loops may involve some new features with respect to the previous ones. Also, the 'content' of the individual steps may differ in relation to the different characteristics of learners in different contexts.

The integration of teaching and research requires interactions between teachers and learners and, therefore, it requires active engagement from both sides. Options involving active engagement enhance the quality of learning (Pinto 2007) and involve research inherently (by their nature). Then, teaching and research mutually integrate within the normal classroom activities. In-class interactions and the analysis of students' works (easily integrated into the works' grading) are the most valuable sources of information, enabling diagnoses and simultaneously providing materials for new in-class activities aimed at enhancing understanding (e.g., through the analysis of errors (Mammino 1996a, 2002a; Love and Mammino 1997). The reliability of these sources is the highest, as they correspond to students' greatest efforts to produce their best. The analysis of students' works is also the most straightforward source for the identification of recurrent aspects. Errors are recurrent when many students make them, and also when the same student repeats them in various contexts. Recurrent errors highlight difficulties in understanding specific

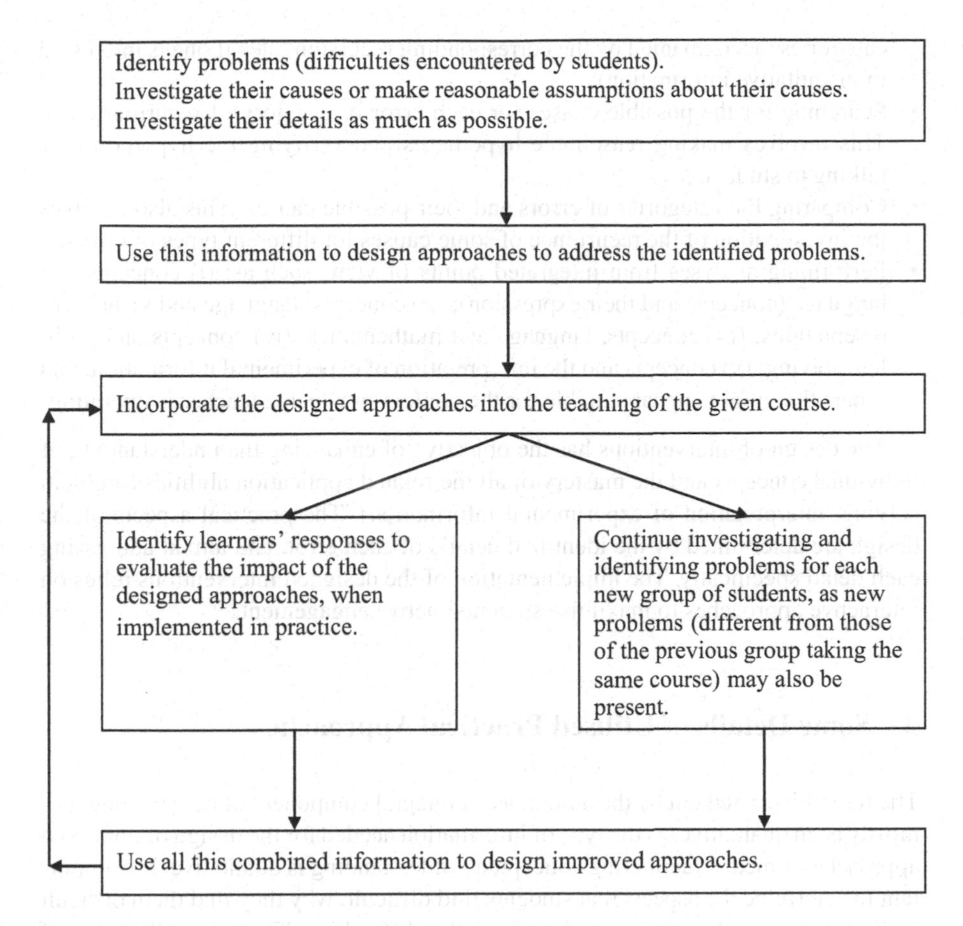

Fig. 10.1 The recursion loop of action research, applied to a course that a teacher teaches every year

concepts, skill inadequacies (e.g., in relation to language, mathematics, or visualization abilities) and previously-generated misconceptions. Specific difficulties are also highlighted by students' questions, or may surface from their answers to verbal or written in-class questions (Beall 1991, 1994; Cooper 1993; Mammino 2006a, 2013a). For chemistry courses, questions in which students are asked to draw images prove particularly useful to diagnose their perceptions about several aspects of the microscopic world (Mammino 1999a, 2002b).

The interpretation of diagnoses requires a plurality of considerations:

- Identifying the details and features of each imprecision or error;
- Verifying and analysing the recurrence rate of specific errors. This involves the recognition of similarities, and of the extent to which some differences may be considered minor enough to enable those errors to be grouped under the same category. The outcome is a classification of errors into a reasonable set of

categories (accompanied by the corresponding recursion rates if one is interested in quantitative information);
- Searching for the possible causes for each error or category of recurrent error. This involves making reasonable hypotheses and verifying the hypotheses by talking to students;
- Comparing the categories of errors and their possible causes. This also involves the investigation of the recurrence of some causes for different types of errors.
- Performing analyses from integrated points of view, such as: (i) concepts and language (concepts and their expression); (ii) concepts, language and visual representations; (iii) concepts, language and mathematics; (iv) concepts and problem solving; (v) concepts and the interpretation of experimental information; and others that might appear suitable for the context in which a teacher is operating.

The design of interventions has the objective of enhancing the understanding of individual concepts and the mastery of all the related application abilities (problem solving, interpretation of experimental information). The practical aspects of the design are determined by the identified details of each error, and aim at addressing each detail specifically. The implementation of the designed interventions relies on interactive approaches to maximise students' active engagement.

3 Some Details of Utilised Practical Approaches

The research carried out by the author, as an integral component of her teaching, has mostly been qualitative, as the type of information needed for the design of improved approaches aimed at facilitating conceptual understanding is qualitative. It is important to determine the aspects that students find difficult, why they find them difficult and what can be done to try and address the difficulties. These are all pieces of qualitative information.

As already mentioned, the only sources of information utilised were in-class interactions and the analysis of students' works, occasionally complemented by personal interviews (when deemed expedient). These are the most reliable sources of information, because they correspond to students' genuine attempts to produce their best. Questionnaires were avoided in the last 20 years, for two main reasons:

- Ethical considerations, prompted by an episode in which a group of students in an Italian university refused to answer questionnaires that were not part of their assessment. Those students explained that they did not find it appropriate that they should use part of their time for something that was of interest only to their lecturer. Although this might entail deeper reflections as a general issue, a strict interpretation of ethics was considered preferable by the author.
- The recollection of the author's and her friends' juvenile (in their students' years) attitude towards the rare questionnaires that were proposed to them – an attitude based on the assumption that something that is not part of the assessment can be

an interesting occasion for fun. Even when students do not 'have fun' explicitly with a questionnaire, their commitment is not likely to equal the one they devote to assessment 'items', so, the information that can be derived from them may be partial or not sufficiently focused.

Ethical considerations also prevented any 'experiment' in which a group of students is taught in one way (e.g., the newly-designed approach) and another group in a different way (e.g., the previous year approach). This is based on the conviction that all the students taking a course are entitled to the best that their teacher can offer in that course. The admissible reference to evaluate the advantages or disadvantages of a new approach is provided by the results of the previous year/s, when that approach was not yet utilised.

Observations-based design may involve major routes meant for a large-scale scope, such as writing textbooks aimed at addressing identified problems point-by-point, or may involve strategic approaches meant to address the problems of a specific course, both for the students' group taking it at a certain moment and by accumulating information and pursuing improvement during subsequent teaching of the same course in subsequent years. The author had an experience of the former type, initiated by analysing the details of the difficulties Italian secondary school students encountered in their first approach to chemistry (during several years, in informal contacts) and then preparing a textbook and some auxiliary books aimed at addressing the identified details one by one (Mammino 1993, 1994, 1998a, 2003a), including a guide to correct expression in the sciences (Mammino 1995a). Most of her research has however been linked to the courses that she has been teaching, and the rest of this article focuses on it. It is important to recall that the content of the next sections is an overview of the author's research and approaches. Several of the issues mentioned in this overview have been objects of investigation by a number of researchers and from various points of view. However, the present overview cannot treat each issue broadly, because doing so would correspond to as many review-articles as the considered issues, and this would by far go beyond the scope of the current chapter.

4 Diagnoses and Categories

This section provides an overview of the main categories of errors, learning difficulties, and their ascertained or probable causes, as identified in the systematic study conducted in the last 30 years within the courses taught by the author. It initially considers few examples for which a more detailed analysis of diagnoses and interventions is outlined (Sect. 4.1) and then proceeds with a rapid recall of other major categories.

4.1 Illustrative Examples of Diagnoses and Interventions

This subsection considers some illustrative examples in terms of what can be regarded as the main 'key' in the diagnosis of a problem, and outlines corresponding interventions. The 'key' is the basis for the classification of errors into categories.

The main 'key' may be identified as incorrect information provided at some stages of the learner's instruction. Misconceptions are thus generated and result in recurrent errors concerning basic chemistry. These types of errors (errors due to inherited incorrect information) require particularly accurate and repeated interventions to induce students to abandon conceptions that they have been taught and have internalised – often with the great depth with which the first-encountered information on a certain issue is internalised. An example is offered by the frequent incorrect answers on the formation reactions of ionic compounds, such as

$$\textit{Formation reaction of NaCl} \quad Na^{+} + Cl^{-} \rightarrow NaCl$$

Interventions need to recall (and thoroughly discuss):

- the definition of formation reaction (a reaction starting from the elements to give 1 mol of a compound);
- the fact that elements (in their elementary state) do not exist as ions, but only as neutral species (they can be ions only in compounds). It would not be possible to have only positive ions or only negative ions in the sample of an element (as would be implied by the use of Na^+ and Cl^- as reactants in a formation reaction).

Then, students are guided to write the correct equation for the formation reaction. Finally, it is recalled that the equation $Na^+ + Cl^- \rightarrow NaCl$ has a meaning only if it is preceded by the equations for the formation of the ions from the neutral atoms, and this, in turn, has a meaning only as part of our models in the explanation of the formation of ionic bonds, not as formation reaction of a compound.

Another frequent error largely due to inherited incorrect information concerns the phenomena occurring when a substance is heated. Many students (often the majority, even in the second year thermodynamics course) list them as (Mammino 2000a, 2008a):

I. *heating of the solid up to the melting point*
II. *melting of the solid*
III. *heating of the liquid up to the boiling point*
IV. *boiling of the liquid*
V. *vaporization*

The interventions invite students to explain the meaning of their answers, in particular, the meaning of considering a "vaporization" transition after a liquid has boiled. They are also invited to mention concrete examples to illustrate their intended meaning. The discussion proceeds until a correct conclusion is reached.

The main 'key' may involve a combination of incorrect information inherited from previous instruction and lack of attention to the meaning (or implications) of everyday life experience (such lack of attention being, in turn, often part of inherited attitudes, because of the inadequacy or absence of references to everyday life in association with taught concepts). This leads to recurrent errors stating things that are in contradiction with aspects of everyday experience, such as the following statement:

In a solid, molecules are very close to each other; in the liquid, they move further away from each other.

The intervention makes reference to everyday experience by asking students to describe what happens if a bottle of water is put in the deep freezer and, after they answer that it cracks, asking whether this is consistent with the idea that molecules are always further apart in the liquid than in the solid. After this, the last step of the intervention focuses on the identification of the features that are always different in the solid state and in the liquid state (for the molecular level, the ordered structure in the solid and its absence in the liquid). The information that glass is a liquid by its molecular structure may contribute to stress the concept that the fundamental difference concerns the arrangement of molecules in the two states, because this information comes as novelty to students and appears to be contradictory to everyday, macroscopic-restricted experience.

Mathematics and everyday life evidence may integrate into main 'keys'. It is the case of recurrent errors involving incorrect mathematics and leading to outcomes in contradiction with everyday experience, as in the following statement:

In the experiment of heating water, we found that time is directly proportional to temperature.

The interventions recall the concept of dependent and independent variables, and stimulate students to consider whether time can be a dependent variable (which is tantamount to consider whether we (as human beings or experimenters) can control its flow); reflections on the one-direction and independent flow of time closely relate to everyday experience.

4.2 Identification of Categories in Relation to Concepts

Difficulties and recurrent errors are encountered in relation to most chemistry concepts, and they extend to many or most of the aspects and implications associated with each concept. It is important to address them as comprehensively as possible, where the 'comprehensiveness' may also involve the utilisation of information from other courses to ensure solid foundations (Mammino 2005a). Even a brief outline of the challenges students encounter with each concept would require much more space than that of a chapter. The next paragraphs provide some illustrative examples in relation to fundamental concepts such as chemical reactions, phase transitions, and atomic orbitals.

An example of the many aspects related to a concept, and the corresponding challenges, is offered by chemical reactions. The concept itself is often not clear. Students may be unable to provide a definition of the term, or an explanation of what a reaction implies, even up to the third year level (Mammino 2004a, 2015a). This hampers the understanding of relevant components of other courses, including typically application-linked courses such as process technology (which, at UNIVEN, is a third year course). The difficulties diagnosed in this course with regard to conservation and non-conservation aspects in a chemical reaction suggested the opportunity of putting great emphasis on them in the first year general chemistry course (Mammino 2005a). Unclear internalisation of the fact that a chemical reaction involves a change in the substances present (a transformation of some substances into others) leads to confusions about the identification of when one can speak of reaction; then, the term *reaction* is used as synonym of *process* (Mammino 2004b), although not all processes are chemical reactions (e.g., phase transitions are processes, but they are not chemical reactions). This distinction is fundamental for problem solving, including plant design; e.g., reactive systems and non-reactive systems are treated differently in the material or energy balances for chemical plants. Difficulties are also encountered for most concepts associated with chemical reactions, such as the balancing of the equations representing them (where the problem is mostly practical), their thermodynamic aspects (Mammino 2013b, 2014a), their spontaneity or non-spontaneity (Mammino 2012a, b), and their kinetics (Mammino 2015b). When balancing is not clearly internalised as being related to atoms and their conservation during a chemical reaction, it is perceived as a sort of mysterious mathematical game disconnected from its physical meaning; this disconnection is the main source of errors, as only the ability of ‘counting’ atoms leads to correct balancing. If a non-sufficiently rigorous statement of the Hess law is provided, students may confuse the imagined steps utilised in its application with the actual mechanism of a given reaction (Mammino 2013b). The definition of spontaneity is easily memorised, but handling the concept with regard to chemical reactions appears to be challenging, also in relation to the fact that the criterion for spontaneity is related to ΔG (free energy change) and not to ΔH (enthalpy change), whereas most students tend to consider a reaction to be spontaneous if it is exothermic (Mammino 2012a). The importance of the spontaneity concept both in thermodynamics and in electrochemistry (where it determines the distinction between galvanic cells and electrolytic cells, (Mammino 2012b)) stresses the importance of clear understanding. For redox reactions, the spontaneity concept relates to the concept of the relative tendencies of elements to be in an oxidised state; students experience difficulties both in understanding this concept (above all its *relative* nature, for which the concept cannot be referred to one element at a time) and in applying it in problem solving, or in the interpretation of qualitative experiments (Mammino 2008b, 2011a).

Phase transitions are encountered as a relevant topic in the first year general chemistry course and in the second year physical chemistry (chemical thermodynamics) course. Knowledge of their relevant features is required within many other topics in several courses. Students experience difficulties with many of these features (Mammino 2000a, 2008a), starting from basic concepts such as the fact that each transition occurs at different temperatures for different substances and that, for the same substance, the temperature of a given transition depends on pressure (for instance, they fail to realize that the fact that the boiling point of water that they measure in their lab is lower than 100 °C is due to the altitude of the place, and try to ascribe it to various – often imaginary – experimental errors). The difficulties also include the interpretation of the heating curve of a substance (as also illustrated in Sect. 4.1) and the macroscopic and microscopic descriptions of each transition.

Atomic orbitals and electronic configurations pose considerable challenges both in the first year (when they are introduced in the general chemistry course) and throughout subsequent years, including postgraduate courses (Mammino 2008b, 2013c). An inherent pedagogical difficulty is the practical impossibility of designing adequate visualizations for electronic configuration. Visualization is limited to the shapes of the individual orbitals and to largely abstract energy levels diagrams. It is not easy to visualise an electron configuration in terms of the shapes of all the orbitals that are occupied in an atom, viewing them simultaneously to give an idea of the situation in that atom. Furthermore, excessively simplified images encountered in previous instruction, combined with the provision of imprecise information, generate misconceptions for which orbitals are perceived as actual boxes (because they are often represented as boxes in energy levels diagrams) where the electrons can find accommodation; this leads to a mental image of the atom not much dissimilar from that of a beehive, with the electrons inhabiting the cells. Counteracting this image is not easy, because of the dominant impact of the first information encountered on a certain issue and of the correspondingly generated mental images. Consequently, concepts utilising orbitals become particularly challenging. Understanding the hybridization of atomic orbitals and its relationships to molecular geometry is challenging during the first year and remains challenging throughout all the other courses, including postgraduate ones (Mammino 2003b).

4.3 *Searching for Causes and Identifying Categories of Background Challenges*

Searching for the causes of the difficulties experienced by students may lead to the identification of generalised inadequacies, such as poor language mastery (whose impact is particularly heavy in second-language instruction contexts) or poor mastery of communication tools such as visualization. The growing awareness that inadequate language mastery is the major cause behind all the other difficulties prompted specific investigation of its major features:

- The analysis of how grammar difficulties impact on the understanding of chemistry concepts or their correct expression (Mammino 1998b, 2005b, 2006b, 2007, 2009a);
- The disadvantages inherent in second-language instruction and the importance of mother tongue instruction (Mammino 2004c, 2010a, 2013d);
- The importance of educating students in the use of the 'language of science' (Mammino 2010b) and of doing so within chemistry courses, and the main features of this endeavour, namely: (i) the recognition of the interdisciplinary nature of the endeavour (Mammino 1995b) and of the favourable opportunities offered by the nature of chemistry, with its diverse components and the importance of descriptions (Mammino 1999b); (ii) the general features of this education (Mammino 2000b, 2010c, 2014b); (iii) for second-language contexts, the importance of giving attention to the characteristics and internal logic of the mother tongue as the familiar "mental home" (Seepe 2000) facilitating better awareness of the meaning actually conveyed by a chosen wording (Mammino 2000c), and the general importance of developing language mastery and narrative abilities within the mother tongue, as the best ground within which to foster abilities that will then be transferable to any other language that a person uses (Mammino 2015c).
- The potentialities of language as a pedagogical resource for chemistry teaching (Mammino 2001a), encompassing the role of precision and rigour as tools to favour understanding-clarity and to prevent confusions, misunderstandings and misconceptions (Mammino 1996b, 1997a, 1998c, 2000d) (including the rigorous use of the words of common language (Mammino 1996c, 2000d, 2006c)); the utilisation of language analysis as a tool for the clarification of chemistry concepts (Mammino 2015d); and the incorporation of attention to language as an integral part of chemistry courses meant to bridge secondary and tertiary instruction (Mammino 2012c).

The design of in-class interventions requires the identification of challenges categories which are often related to general background difficulties (such as language mastery inadequacies) and need to be singled-out for specific interventions. Several such categories have been identified. From the way in which these categories have been termed during the study, they may appear to have a philosophical nature (and actually they do have it), but the pedagogical implications are highly practical, as they provide indications about what interventions need to focus on. The interventions are then incorporated as integral components of the teaching activity. Relevant categories of challenges are mentioned in the following paragraphs, together with the types of interventions that have gradually been incorporated into the presentation of new topics to students.

The interplay between microscopic and macroscopic descriptions in chemistry (Mammino 1997b; Mammino and Cardellini 2005) is a challenge acknowledged by chemistry educators worldwide (e.g. Ben-Zvi et al. 1986; De Posada 1997; Gabel et al. 1987; Harrison and Treagust 1996; Lijnse et al. 1990; Maskill et al. 1997; Mitchell and Kellington 1982). Interventions involve frequent recalling to what pertains to the microscopic description and what pertains to the macroscopic description, with references to the features relevant for each issue. For instance, when introducing the mole concept, both the mass of a molecule and the mass of a mole of the same substance are shown in grams, writing the mass of a molecule with all the needed zeros before the first non-zero decimal figure to provide a visual impact of the enormous difference between the two masses.

The distinction between statements or information with general characters and those valid only for specific (particular) cases is fundamental for conceptual clarity. Unclear distinction leads to errors, such as the consideration that all substances melt at 0 °C and boil at 100 °C (Mammino 2000a, 2001b, 2009b), or the use of equations derived for particular cases as if they had general validity. The interventions involve underlining what has general validity, and what does not, both during explanations and in the form of in-class questions.

Cause-effect relationships have fundamental roles in all the scientific discourse, as they constitute a major guideline for the interpretation of observations and experimental results. Poor language mastery often results in unclear identification of the cause and the effect in a given situation (Mammino 2006d). Interventions involve an introductory explanation of their meaning at the first opportunity within a course (with the utilization of examples from everyday life), and clear recalling of what is the cause and what is the effect during explanations of new phenomena for whose understanding the distinction is important.

Unclear distinction between systems and processes affects the description of either (Mammino 2002c), both in terms of word association (e.g. using the verbs *occur* or *happen* in relation to systems, while they are suitable only for processes) and in terms of understanding-clarity. Confusions between physical quantities and their changes affect the understanding level and also the problem solving level (Mammino 2001c). In both cases, it is important to underline the distinction, explicitly attracting students' attention to it. It is particularly relevant to do so in the chemical thermodynamics (physical chemistry) course, where understanding the characteristics of a system, and of the process or processes occurring in it, is fundamental, and where confusions between U and ΔU, H and ΔH, S and ΔS etc. result in serious errors when solving problems.

Other distinctions play key roles for clear understanding. Unclear distinction between numbers and values (Mammino 2003c) often results in unclear understanding of the difference between objects (which can be numbered) and physical quantities (which are measured). Then students use incorrect wordings such as "number of mass" in relation to the mass of a sample, or "number of concentration" in relation to the concentration of a solution.

Aspects such as the above-mentioned understanding of distinctions between entities of different nature, or the recognition of logical categories (such as the cause-effect relationship), are associated with the basics of the scientific method; the ability to express them requires understanding of these basics. Inadequate language mastery is largely responsible for the diagnosed confusions, as well as for the inadequacies in the inherent logic of individual answers or descriptions (e.g., in reports) (Mammino 2010a). It is also responsible for other inadequacies concerning method-related aspects, such as unclear understanding of the operational role of definitions and their inherent nature as statements providing all the information that is needed to recognise the item that is being defined (Mammino 2000e). Unclear recognition of this operational role results in the removal of relevant pieces of information from definitions, or in the provision of statements that are not consistent with that role. An example is the statement saying that the oxidation number is a "value which is assigned to elements according to a certain set of rules"; this statement has no operational character because, by itself, does not enable the identification of the oxidation number of an element in practice (the "set of rules" cannot be included as part of a definition and, therefore, the statement is not a definition).

Inadequate familiarity with basic aspects of the scientific method also hampers the interpretation of experimental information, where features such as cause-effect relationships play fundamental roles both in the interpretation of results and in the identification of possible sources of experimental errors (Mammino 2010d). As a consequence, descriptions and interpretations provided in the reports are too often distant from reality (Mammino 2016). Recurrent misinterpretations appear for experiments requiring greater clarity of conceptual understanding, such as the qualitative experiment on the relative tendencies of selected elements to be in an oxidised state (an experiment in which each selected metal is dipped in solutions containing ions of the other selected metals (Mammino 2011b)).

Visualization plays fundamental roles in the teaching and learning of chemistry, above all (but not only) to illustrate the entities of the microscopic world (molecular structures etc.). Understanding the information conveyed by images requires an adequate degree of visual literacy. The acquisition of visual literacy largely relies on language mastery (Mammino 2010a, e). The combination of poor language mastery and poor visual literacy sharply decreases communication effectiveness, thus hampering learning. Teaching and learning rely on communication. Their effectiveness depends on the extent to which students master communication tools and on the extent to which the teacher can utilise them with sufficient flexibility to try and maximise communication with the given group of students. When students' mastery of the communication tools is poor, communication often fails. Visualisation can provide a type of communication which can integrate and complement the communication through language, but it cannot communicate information on new concepts without the support of language. The teacher can try to cultivate both abilities in an integrated why, by utilising each of them as support for the other, with strategies tailored to the characteristics of the given group of students (Mammino 2014c). The design of images also needs to consider the characteristics of the group of students and the features that make certain groups disadvantaged (Mammino 2005c). Attention to reality and to the scientific method requires that adequate information is also provided about the limitations of visualization, first of all about the fact that the inherent simplifications frequently disguise the actual complexity of the systems considered (e.g., molecular systems (Mammino 2013e)). For instance, it is important to stress that the boxes, circles or lines representing orbitals in energy level diagrams are abstract representations for us to follow energy trends, but do not correspond to any space allocations or space shapes or space partitions within the atoms.

5 Discussion and Conclusions

The aspects, challenges and examples considered in the previous sections are only a fraction of the observations, interpretations and inferences for in-class interventions accumulated during 30 years of integration of teaching and educational research. However, they may suffice to highlight some fundamental aspects. In order to continuously improve its quality, teaching involves a huge amount of investigation, which can be viewed as experimental research focused on the 'real students' a teacher is engaged with. Diagnoses about students' difficulties and their probable causes constitute the basis for the design of specifically-tailored interventions. Students' active engagement is fundamental for the interventions to have an impact. In-class oral and written questions, error analysis (a powerful explanation tool), collaborative construction of images (Mammino 1999a), and other options inherently requiring students' participation, are functional not only to enhance conceptual understanding and the acquisition of practical abilities, but also to address behaviour challenges of specific contexts, such as passive attitudes and the habit of passive memorization. Students' responses to the interventions constitute a new source of information leading to new diagnoses. This «research + implementation» loop is iterative and does not reach a final stage, but aims at continuous improvement.

The combination of observations, analysis, interpretation and design is an aspect that removes any risk of 'boredom' from the teaching activity, keeping it alive and challenging even if one is teaching the same course and the same concepts year after year, because every new year brings a new design on the basis of the observations of the previous years, and new challenges from the new group of students. It also removes any risk of 'boredom' from the grading component, because it is the component that provides the necessary information for a teacher to understand students' difficulties and to design ways to address them, both in the short term (with the same group, in post-test or post-report sessions) and in the long term (for the subsequent years). The design entails the specific challenges of being meant for human beings, with their inherent diversity: how to design approaches that really respond to the actual needs of the students involved, considering that each group of students is made by a number of individuals with different abilities and attitudes, and the design should try to meet the needs of each of them.

Because of being aimed at designing options to address specific problems, the research described here mostly focused on individual issues (challenges with specific concepts or impact of specific background problems, as indicated by most of the references considered in the previous sections). However, it can also expand to comprehensive outlines of the situation in the given context, or to generalized outlines which may be meaningful for chemistry education on the whole (e.g., Mammino 2008c, 2010a, 2011c, 2014d). The integration of teaching and research is suitable and fruitful for all levels of instruction, including specialised courses of the last undergraduate or the early postgraduate years (e.g., Mammino 1995c, 1998d, 2003d, 2004d, 2005d, 2006e, 2015e). It is accessible to any teacher who wishes to continuously enhance the quality of teaching and learning with his/her students by paying careful attention to what students do and how they respond and perform. Furthermore, discussing one's own teaching approaches with the students (why one is using a certain approach and what the expected benefits are) may contribute further insights relevant for diagnoses and design; and some students might develop keen interest in the challenges of education (as it happened to the author, whose secondary school chemistry teacher used to discuss his teaching approaches with the class).

References

Beall, H. (1991). In-class writing in general chemistry: A tool for increasing comprehension and communication. *Journal of Chemical Education, 68*(1), 148–149.

Beall, H. (1994). Probing student misconceptions in thermodynamics with in-class writing. *Journal of Chemical Education, 71*(12), 1056–1958.

Ben-Zvi, R., Eylon, B., & Silberstein, J. (1986). Is an atom of copper malleable? *Journal of Chemical Education, 63*(1), 64–66.

Cooper, M. M. (1993). Writing – An approach for large enrolment chemistry courses. *Journal of Chemical Education, 70*(6), 476–477.

De Posada, J. M. (1997). Conceptions of high school students concerning the internal structure of metals and their electric conduction: Structure and evolution. *Science Education, 81*(4), 445–467.

Gabel, D. L., Samuel, K. V., & Hunn, D. (1987). Understanding the particulate nature of matter. *Journal of Chemical Education, 64*(8), 695–697.

Girod, M. (n.d.). *Mark Girod's professional library and resources*. www.wou.edu/~girodm/library/ch9.pdf

Harrison, A. G., & Treagust, D. F. (1996). Secondary students' mental models of atoms and molecules: Implications for teaching chemistry. *Science Education, 80*(5), 509–534.

Levin, R. (1947). Frontiers in group dynamics. II. Channels of group life: Social planning and action research. *Human Relations, 1*, 143–159.

Lijnse, P. L., Licht, P., Vos, W., & Waarlo, A. J. (1990). *Relating macroscopic phenomena to microscopic particles. A central problem in secondary science education*. Utrecht: CD-β Press.

Love, A., & Mammino, L. (1997). Using the analysis of errors to improve students' expression in the sciences. *Zimbabwe Journal of Educational Research, 9*(1), 1–17.

Mammino, L. (1993). *Chimica Viva*. Florence: G. d'Anna.

Mammino, L. (1994). *Industria, Ambiente, Evoluzione storica*. Florence: G. d'Anna.

Mammino, L. (1995a). *Il linguaggio e la scienza (guida all'espressione corretta nelle scienze)*. Turin: Società Editrice Internazionale.

Mammino, L. (1995b). L'educazione al linguaggio della scienza: un problema di interdisciplinaritá. *Nuova Secondaria, 5*, 84–86.

Mammino, L. (1995c). Teaching/learning theoretical chemistry at undergraduate level. *Southern Africa Journal of Mathematics and Science Education, 2*(1&2), 69–88.

Mammino, L. (1996a). L'analisi degli errori come strumento didattico. *Nuova Secondaria, 10*, 75–76.

Mammino, L. (1996b). Algunas reflexiones sobre el rigor conceptual y formal en al enseñanza de la química. *Anuario Latinoamericano de Educación Química, 8*(2), 375–384.

Mammino, L. (1996c). Parole comuni nel linguaggio scientifico. *Nuova Secondaria, 6*, 77–77.

Mammino, L. (1997a). Il rigore: esigenza del linguaggio della scienza e strumento didattico. *Scuola Viva, 8*, 14–17.

Mammino, L. (1997b). Estudio de la forma en que los alumnos perciben la distinción y las mutuas relaciones entre el mundo microscópico y el mundo macroscópico. *Anuario Latinoamericano de Educación Química, 9*(2), 178–189.
Mammino, L. (1998a). *Glossario di Chimica*. Florence: G. d'Anna.
Mammino, L. (1998b). Science students and the language problem: Suggestions for a systematic approach. *Zimbabwe Journal of Educational Research, 10*(3), 189–209.
Mammino, L. (1998c). Precision in wording: A tool to facilitate the understanding of chemistry. *CHEMEDA, the Australian Journal of Chemical Education,* 48 & 49 & 50, 30–38.
Mammino, L. (1998d). Enseñanza interactiva en cursos de química teórica. *Anuario Latinoamericano de Educación Química, XI*, 275–278.
Mammino, L. (1999a). Explorando el empleo de las imagines como instrumento de enseñanza interactiva. *Anuario Latinoamericano de Educación Química, XII*, 70–73.
Mammino, L. (1999b). La chimica come area ideale per l'educazione al linguaggio della scienza. In *Edichem '99: Chemistry in the perspective of the new century* (pp. 212–230). Bari: Societa' Chimica Italiana.
Mammino, L. (2000a). Alternative conceptions and misconceptions about the phases of matter. In *SAARMSTE conference proceedings*, pp. 284–289.
Mammino, L. (2000b). Educating towards the language of science. In S. Seepe & D. Dowling (Eds.), *The language of science* (pp. 72–93). Johannesburg: Vyvlia Publishers.
Mammino, L. (2000c). Studying the details of the transition from the mother tongue to the second language. In S. Seepe & D. Dowling (Eds.), *The language of science* (pp. 94–101). Johannesburg: Vyvlia Publishers.
Mammino, L. (2000d). Rigour as a pedagogical tool. In S. Seepe & D. Dowling (Eds.), *The language of science* (pp. 52–71). Johannesburg: Vyvlia Publishers.
Mammino, L. (2000e). Il ruolo delle definizioni nell'insegnamento delle scienze. *Scuola e Cittá, 1*, 25–37.
Mammino, L. (2001a). Using the resources offered by the «language of science» to facilitate the familiarisation with science and technology. In *Proceedings of the CASTME-UNESCO-HBCSE conference on science, technology and mathematics education for human development*, Vol. II, pp. 368–372.
Mammino, L. (2001b). *General* y *particular* en química. *Anuario Latinoamericano de Educación Química, XIV*, 25–28.
Mammino, L. (2001c). Physical quantities and their changes. Difficulties and perceptions by chemistry students. *SAARMSTE (Southern African Association for Research in Mathematics, Science and Technology Education) Journal, 5*, 29–40.
Mammino, L. (2002a). Empleo del análisis de errores para aclarar conceptos de química general. *Enseñanza de las Ciencias, 20*(1), 167–173.
Mammino, L. (2002b). Imagery: A tool for the generation, expression and recognition of students' views about the world of molecules. In *SAARMSTE conference proceedings*, pp. 237–243.
Mammino, L. (2002c). La percepción de la distinción entre sistemas y procesos por parte de los alumnos de química. *Anuario Latinoamericano de Educación Química, XV*, 125–129.
Mammino, L. (2003a). *Chimica Aperta*. Florence: G. d'Anna.
Mammino, L. (2003b). La comprensión de la hibridación de los orbitales atómicos entre aspectos conceptuales y aspectos de lenguaje. *Anuario Latinoamericano de Educación Química, XVI*, 251–256.
Mammino, L. (2003c). La actitud cerca de números y valores en cursos de química general. *Anuario Latinoamericano de Educación Química, XVI*, 5–10.
Mammino, L. (2003d). Addressing the abstractness perception in theoretical chemistry courses. *Journal of Molecular Structure (THEOCHEM), 621*, 27–36.
Mammino, L. (2004a). Percepciones de alumnos de química acerca del fenómeno «reacción química». *Anuario Latinoamericano de Educación Química, XVII*, 42–46.

Mammino, L. (2004b). Dominancia del concepto de reacción, y de los términos asociados, en las percepciones de alumnos de química. *Anuario Latinoamericano de Educación Química, XVIII*, 46–52.
Mammino, L. (2004c). L'impatto, sull'apprendimento della chimica, dell'uso di una lingua diversa da quella materna. In P. Fetto (Ed.), *Atti IV Conferenza Nazionale sull'insegnamento della Chimica* (pp. 117–129). Perugia: Societa' Chimica Italiana.
Mammino, L. (2004d). Mentioning fuzzy logic in theoretical chemistry courses: Motivations and extent. *Journal of Molecular Structure (THEOCHEM), 709*, 231–238.
Mammino, L. (2005a). Suggestions for the enhancement of first year chemistry teaching, from a process technology course. *ICUC Quarterly, 1*(2), 9.
Mammino, L. (2005b). Language-related difficulties in science learning. I. Motivations and approaches for a systematic study. *Journal of Educational Studies, 4*(1), 36–41.
Mammino, L. (2005c). Visualización de conceptos químicos en contextos en desventaja. *Anuario Latinoamericano de Educación Química, XIX*, 177–181.
Mammino, L. (2005d). Method-related aspects in an introductory theoretical chemistry course. *Journal of Molecular Structure (THEOCHEM), 729*, 39–45.
Mammino, L. (2006a). The use of questions in the chemistry classroom: An interaction instrument with maieutic nature. *Anuario Latinoamericano de Educación Química, XXI*, 241–245.
Mammino, L. (2006b). Language-related difficulties in science learning. II. The sound-concept correspondence in a second language. *Journal of Educational Studies, 5*(2), 189–213.
Mammino, L. (2006c). *Terminology in science and technology – An overview through history and options*. Thohoyandou: Ditlou Publishers.
Mammino, L. (2006d). Cause and effect – A relationship whose nature involves conceptual understanding, scientific method, logic and language. *Anuario Latinoamericano de Educación Química, XXI*, 33–37.
Mammino, L. (2006e). The recent history of theoretical chemistry presented from a method-related perspective. *Journal of Molecular Structure (THEOCHEM), 769*(1–3), 61–68.
Mammino, L. (2007). Language-related difficulties in science learning. III. Selection and combination of individual words. *Journal of Educational Studies, 6*(2), 199–214.
Mammino, L. (2008a). Chemistry students' perceptions of the factors influencing phase transitions of pure substances, as expressed in laboratory reports. *Anuario Latinoamericano de Educación Química, XXIII*, 96–100.
Mammino, L. (2008b). Orbitals in chemistry and in chemical education. *Anuario Latinoamericano de Educación Química, XXIII*, 42–46.
Mammino, L. (2008c). Teaching chemistry with and without external representations in professional environments with limited resources. In J. K. Gilbert, M. Reiner, & M. Nakhlekh (Eds.), *Visualization: Theory and practice in science education* (pp. 155–185). Dordrecht: Springer.
Mammino, L. (2009a). Language-related difficulties in science learning. IV. The use of prepositions and the expression of related functions. *Journal of Educational Studies, 8*(4), 142–157.
Mammino, L. (2009b). Teaching physical chemistry in disadvantaged contexts: Challenges, strategies and responses. In M. Gupta-Bhowon, S. Jhaumeer-Laulloo, H. Li Kam Wah, & P. Ramasami (Eds.), *Chemistry education in the ICT age* (pp. 197–223). Dordrecht: Springer Netherlands.
Mammino, L. (2010a). The mother tongue as a fundamental key to the mastering of chemistry language. In C. Flener & P. Kelter (Eds.), *Chemistry as a second language: Chemical education in a globalized society* (pp. 7–42). Washington, DC: American Chemical Society.
Mammino, L. (2010b). The essential role of language mastering in science and technology education. *International Journal of Education and Information Technologies, 3*(4), 139–148.
Mammino, L. (2010c). Language aspects in science and technology education: Novel approaches for new technologies. In P. Dondon & O. Martin (Eds.), *Latest trends on engineering education* (pp. 480–486). Greece: WSEAS Press.

Mammino, L. (2010d). Familiarity with the basics of the scientific method as a prerequisite to identifying the causes of experimental errors. *Anuario Latinoamericano de Educación Química, XXV*, 167–172.
Mammino, L. (2010e). Interplay, interfaces and interdependence between visual literacy and language mastering, as highlighted by chemistry students' works. *Anuario Latinoamericano de Educación Química, XXV*, 16–20.
Mammino, L. (2011a). The relative tendencies of elements to be in an oxidised state. I. Students' understanding of the background concepts. *Anuario Latinoamericano de Educación Química, XXVI*, 59–64.
Mammino, L. (2011b). The relative tendencies of elements to be in an oxidised state. II. Students' attitudes toward qualitative experimental information. *Anuario Latinoamericano de Educación Química, XXVI*, 104–109.
Mammino, L. (2011c). Teaching chemistry in a historically disadvantaged context: Experiences, challenges, and inferences. *Journal of Chemical Education, 88*(11), 1451–1453.
Mammino, L. (2012a). The spontaneity concept: An investigation of the dichotomy between learning the definition and handling the concept. *Anuario Latinoamericano de Educación Química, XXVII*, 120–125.
Mammino, L. (2012b). An investigation of students' difficulties in handling the spontaneity concept in electrochemistry. *Anuario Latinoamericano de Educación Química, XXVII*, 155–160.
Mammino, L. (2012c). Focused language training as a major key for bridging the gap between secondary and tertiary instruction. In D. Mogari, A. Mji, & U. I. Ogbonnaya (Eds.), *ISTE international conference proceedings* (pp. 278–290). Pretoria: UNISA Press.
Mammino, L. (2013a). Teacher-students interactions: The roles of in-class written questions. In M.-H. Chiu (Ed.), *Chemistry education and sustainability in the global age* (pp. 35–48). Dordrecht: Springer.
Mammino, L. (2013b). The Hess law – Chemical information from a rigorous statement and challenges experienced by students. *Anuario Latinoamericano de Educación Química, XXVIII*, 167–172.
Mammino, L. (2013c). Electrons and orbitals: Challenges at first year level and beyond. In D. Mogari, A. Mji, & U. I. Ogbonnaya (Eds.), *ISTE international conference proceedings* (pp. 133–147). Pretoria: UNISA Press.
Mammino, L. (2013d). Importance of language mastery and mother tongue instruction in chemistry learning. In Z. Desai, M. Qorro, & B. Brock-Utne (Eds.), *The role of language in teaching and learning science and mathematics* (pp. 33–56). Somerset West: African Minds.
Mammino, L. (2013e). Stimulating the awareness of complexity while utilizing visualizations of molecules in chemical education. *Anuario Latinoamericano de Educación Química, XXVIII*, 132–137.
Mammino, L. (2014a). The Hess law and the state-function nature of enthalpy. The concept and its expression by chemistry students. *Anuario Latinoamericano de Educación Química, XXIX*, 100–105.
Mammino, L. (2014b). Essential roles of language mastery for conceptual understanding and implications for science education policies. In T. Marek, W. Karwowski, M. Frankowitz, J. Kantola, & P. Zgaga (Eds.), *Human factors of a global society: A system of systems perspective* (pp. 835–847). Rosa Boca: Taylor & Francis Inc, CRC Press.
Mammino, L. (2014c). The interplay between language and visualization: The role of the teacher. In B. Eilam & J. Gilbert (Eds.), *Science teachers' use of visual representations* (pp. 195–225). Dordrecht: Springer.
Mammino, L. (2014d). The first year general chemistry course: Great challenges and great potentialities. In D. Mogari, U. I. Ogbonnaya, & K. Padayachee (Eds.), *ISTE international conference proceedings* (pp. 40–48). Pretoria: UNISA Press.
Mammino, L. (2015a). What happens during a chemical reaction? Insights into students' perceptions. *Anuario Latinoamericano de Educación Química, XXX*(2), 20–25.

Mammino, L. (2015b). Students' understanding of the rate law, as highlighted by laboratory reports. *Anuario Latinoamericano de Educación Química, XXX*(1), 175–180.
Mammino, L. (2015c). Language mastery, narrative abilities and oral expression abilities in chemistry learning: Importance of mother tongue and traditional narration. *South African Journal of African Languages, 5*(1), 19–27.
Mammino, L. (2015d). Clarifying chemistry concepts through language analysis. In J. Lundell, M. Aksela, & S. Tolppanen (Eds.), *LUMAT Special Issue of ECRICE, 3*(4), pp. 482–500.
Mammino, L. (2015e). Challenges of the quantum chemistry course in an inadequate-epistemological-access context and exploration of addressing options. In L. D. Mogari (Ed.), *ISTE international conference proceedings* (pp. 275–244). Pretoria: UNISA Press.
Mammino, L. (2016). Being consistent with reality: A great challenge for chemistry laboratory reports. In J. Kriek, B. Bantwini, C. Ochonogor, J. J. Dhlamini, & L. Goosen (Eds.), *ISTE conference on mathematics, science and technology education proceedings* (pp. 428–437). Pretoria: UNISA Press.
Mammino, L., & Cardellini, L. (2005). Studying students' understanding of the interplay between the microscopic and the macroscopic descriptions in chemistry. *Baltic Journal of Science Education, 1*(7), 51–62.
Maskill, R., Cachapuz, A. F., & C & Koulaidis V. (1997). Young pupils' ideas about the microscopic nature of matter in three different European countries. *International Journal of Science Education, 19*(6), 631–645.
Mitchell, A. C., & Kellington, S. H. (1982). Learning difficulties associated with the particulate theory of matter in the Scottish integrated science course. *European Journal of Science Education, 4*(4), 429–440.
Pinto, C. G. (Ed.). (2007). *Aprendizaje Activo de la Física y la Química* (Active learning of physics and chemistry). Madrid: Equipo Sirius.
Seepe, S. (2000). A pedagogical justification for mother tongue instruction. In S. Seepe & D. Dowling (Eds.), *The language of science* (pp. 40–51). Johannesburg: Vyvlia Publishers.

Chapter 11
Teaching Modern Physics to Chemistry Students

Joseph K. Kirui and Lordwell Jhamba

1 Introduction

1.1 Curriculum and Background

The course of Modern Physics (PHY2624) is offered to Bachelor-of-Science students in their second year at the University of Venda (UNIVEN). The students registered for chemistry major also take modern physics. The course is taught by lecturers from the physics department as one of their core modules. The chemistry students attend this course together with students majoring in physics.

The students that form the subject of this article must have taken physical science and mathematics at high school and obtained at least a level 4, or 50% and above, in each of them. In addition, the School of Mathematical and Natural Sciences admission requirements must be met as stipulated in the University calendar on a yearly basis.

In the physics section of the physical science grade 12 curriculum at high school, the students have been taught some modern physics topics, such as absorption and scattering of light, photoelectric effect, and the dual nature of light (or wave-particle-duality of light).

In their first year undergraduate curriculum, the students take four compulsory modules in physics, namely: Mechanics (PHY1521), Waves and Optics (PHY1522), Properties of Matter and Heat (PHY1623) and Electricity and Magnetism (PHY1624), all taught by lecturers from the Physics Department.

The co-requisite mathematics modules are Differential Calculus (MAT1541) and Integral Calculus (MAT1641), both taught by lecturers from the Department of Pure and Applied Mathematics. The time allocated for lectures for each of the six

J. K. Kirui (✉) · L. Jhamba
Department of Physics, University of Venda, Thohoyandou, South Africa
e-mail: joseph.kirui@univen.ac.za

L. Mammino, J. Apotheker (eds.), *Research in Chemistry Education*,
https://doi.org/10.1007/978-3-030-59882-2_11

modules is 2 h per week for one semester of 13 weeks. In addition, for each of the four physics modules, there is also a laboratory session of 3 h per week.

As for the methods of assessment of Modern Physics, they comprise the continuous assessment during the semester and a 3-h examination at the end of the semester. Two tests and an assignment administered well before the end of the semester form the continuous assessment or semester mark.

The main teaching aids for delivering lectures consist of a data projector and a whiteboard.

1.2 *The Importance of Modern Physics in Undergraduate Curriculum*

Modern Physics, as taught to the UNIVEN undergraduates, has two main components – special relativity and introductory quantum mechanics. Relativity deals with the notions of space and time, whereas quantum mechanics introduces probabilities as well as uncertainties in positions and momenta into the motion of microscopic particles.

Relativity, Quantum Mechanics and Chaos theory are three of the most significant scientific advances of the twentieth century – each fundamentally changing the way we understand the physical universe (Shabajee and Postlethwaite 2000).

The theory of relativity has enabled us to view the world differently. It is now part of undergraduate curricula in many universities worldwide. In fact, suggestions to include it in high school curricula have been made (Arriassecq and Greca 2012; Villani and Arruda 1998). To the general population, it should be pointed out that one very popular application of the theory of relativity is the Global Positioning System (GPS).

Another application of modern physics is the theory of chaos (Gleick 1988), which enables the modelling of phenomena such as weather patterns, motor traffic, the animal populations, and even the functioning of the human heart.

Quantum mechanics plays an important and central role in modern technologies. Training in the discipline is becoming mandatory for future physicists, chemists, engineers and biologists, whose core subjects at first year level include introductory physics courses (McDermott and Redish 1999; Müller and Weisner 2002; McKagan et al. 2010). In chemistry, a good grounding in quantum mechanics is necessary to solve complex chemical problems and to model molecular structures through novel computational techniques that are founded on the quantum theory.

Because of the reasons just explained, ways must be found to make those areas more accessible to students. The challenge is that the two theories are very abstract in nature. Any research about the challenges students encounter in understanding these theories is of paramount importance in science education at the present moment.

1.3 The Peculiarity of Quantum Physics and Challenges to Science Education

The difficulties in learning quantum physics are manifold and widespread for both advanced and introductory courses (Fishler and Lichtfeldt 1992; Tsaparlis 1997; Johnston et al. 1998; Singh 2001; Cataloglou and Robinett 2002; Taber 2002). The same applies to the modern physics course offered to the chemistry students at UNIVEN.

It is important to realize that the problems are not inherent in students' abilities, but must be found in the teaching methodology, the presentation and the interaction between the presenter and the receptor; the lecturer and the learner must find optimum conditions for effective learning of this difficult but unavoidable course. The physics and chemistry education fraternity must rise to the challenge. They must consolidate gains in the educational research and enable implementation of new educational methodologies for a better way forward.

It can no longer be assumed that the learners are not competent. There is more to the challenges of education than the misconceptions arising from the obvious factors of poor background, lack of mathematical ability, poor delivery of lectures and lack of teaching aids. These causes may bedevil small departments like those at UNIVEN, but should not be the case for larger establishments where, nonetheless, research has shown education to be equally wanting in the optimum uptake of the theories of quantum mechanics and relativity (Fishler & Lichtfeldt).

1.4 Common Topics That Students Find Challenging When Learning Modern Physics

Although the authors have not designed and implemented a specific research instrument to investigate the challenges faced by the chemistry students and their physics counterparts, several very common sticking points were noted both during lecture sessions and in the assessment of assignments and examinations. The most common of the observed conceptual difficulties usually fall under the topics considered in the next paragraphs.

Simultaneity and Relativity of Space Students of modern physics learned about length contraction, though it was not easy for them to extrapolate this information to facilitate the understanding of the non-absoluteness of space.

The Doppler Effect Students generally find it odd that the Doppler effect is treated differently for sound and for light. This is because for sound, two velocities are required – for the source and the detector with respect to the air. However, on accepting the postulates of special relativity, most of the students now appreciate that light does not need any medium to transmit – hence only the relative velocity is required. But accepting the postulates is another big challenge for students.

The Speed of Light Chemistry students also take a course in electrodynamics in the second semester. The topic on special relativity comes very early in the modern physics course also in the second semester. The result is that many students do not fully appreciate the power of Maxwell's equations, as these are offered much later in the electrodynamics course. The study of the theories of Rayleigh and Jeans enable the students to appreciate that, in science, models are developed in such a way that they fit experimental data.

Wave-Particle Duality The topic of wave-particle duality is very abstract to most students. The students find it difficult to comprehend waves of matter. After a first year course on waves and optics, they think they know about the properties of waves. Hence they ask what is undergoing displacements in order to register amplitudes for the matter waves. An effective way to make the students appreciate matter waves is the discussion of the experiment on electron diffraction, firstly done by Davisson and Germer (1928).

The Schrödinger Equation and Quantum Systems The fact that the Schrödinger equation cannot be derived from the principles of classical mechanics complicates its understanding by students encountering it for the first time. Lecturers have to emphasize and instil the importance of the role of the Schrödinger equation, namely, that its solutions provide full descriptions of atoms and molecules and are in agreement with atomic and nuclear physics experiments.

2 Conceptual Challenges in Learning Modern Physics Encountered by Chemistry Students

2.1 Perceived Misconceptions and Challenges Faced by Students

A large body of research has documented students' conceptions and learning difficulties within a range of topics in classical physics (Duit 2009); however, very little documentation is available on research on the understanding and learning of modern physics. The main challenges for understanding quantum physics and relativity stem from the fact that they concern phenomena that cannot be visualized or experienced directly; and it is not invariably easy to perform classroom experiments to illustrate the phenomena or aid understanding.

The works by Pitts et al. (2014) are among the studies aimed at unravelling the challenges facing students of modern physics. These researchers found that students defined and described a frame of reference, but believed that there is a preferred special frame for an observer (Arriasecq and Greca 2012). Similarly, students could not fully understand that two events which may be simultaneous in one frame of reference are not necessarily simultaneous in another inertial reference frame. In addition, they held the belief that both time dilation and length contraction do not represent a reality (Scherr 2007; Scherr et al. 2001; Hewson 1982; Posner et al. 1982).

2.2 Is It Misconceptions or Alternative Conception?

There is a constructivist theory about learning which presents the view that knowledge cannot be transmitted, but must be constructed by the mental activity of students. This is a perspective in science education, which has excited the interest of educational researchers and found substantial appeal (Driver et al. 1994). According to this view, the challenges that we understand as students' misconceptions are given another interpretation: they are thought to actually be alternative conceptions, also termed preconceptions or mind models (Chhabra and Bajeva 2012).

The fact that the alternative conceptions are firmly in the students' minds poses a challenge to the learning of new related concepts (Muller and Sharma 2007). The students are now facing additional problems in learning new ideas correctly, and may not be receptive to any opinion contrary to their preconceptions.

3 Major Factors Contributing to the Learning Challenges

According to the constructivist approach to learning, (Driver et al. 1994), a student

> ….will construct knowledge, communicated and validated within everyday culture.

Since there is bound to be difference between how people reason in a daily interaction and the established scientific format of reasoning, alternative view points on almost any idea become possible. The end result is that the learning of important scientific facts is severely stifled.

A teacher will have personal views about teaching, learning and knowledge. These views determine the way in which he/she teaches modern physics and will have an effect on the learning of this crucial subject (Bencze et al. 2006).

Some of the factors complicating the teaching and learning of Modern Physics at UNIVEN are the following:

- Lack of adequate teaching aids – though a unit called the Centre for Higher Education Teaching and Learning (CHETL) is expected to change this, however slightly.
- heavy teaching loads for the lecturers;
- inadequate tutorial support system;
- inadequate computer facilities;

- poor grasp of the English language for a number of students;
- poor mathematical ability for some students;
- frequent lack of sufficient competence in the basics of probability theory.

4 Possible Strategies and Steps Needed to Address the Teaching and Learning Challenges

The strengthening of physics education at the physics department should be encouraged. It is noted that Chemistry education research is already more active than physics education research at UNIVEN.

The educators should endeavour to understand students' prior knowledge or preconceptions about the more challenging topics in the course. In this way, lecturers will be in a better position to help initiate the much needed paradigm shift from classical physics thinking to quantum mechanical thinking, as well as with the special relativity concepts.

Non-interpretive questions in tests should be minimized, since such questions can be answered by merely committing the correct contents of the course to memory and then regurgitating them without any need to think.

Experiments should be found that are hands-on and enable the students to discover for themselves important quantum mechanical properties involved (Zollman et al. 2002).

Students must be assisted to develop a quantum mechanical worldview, which needs a paradigm shift (Posner et al. 1982). This is probably best achieved by highlighting the differences between quantum mechanics and classical mechanics, and convincing students to change their view radically.

The introduction of computer-based learning can help speed up the uptake of quantum mechanics and modern physics (Singh et al. 2006). Furthermore, well-designed tutorials that take into consideration cognitive issues will help greatly.

5 Conclusions

It is of crucial importance to understand and appreciate students' difficulties in understanding the basic elements of the theory of relativity. This should lead to the construction of an effective educational approach.

It is also important to realize that the mode of teaching may be wanting and efforts must be made to seek and implement the research findings in the international literature, about the teaching and learning of physics.

A system of working together to solve educational problems should be shared between the departments of physics and chemistry, since the difficulties are not confined to the set of students of only one department. Unified approaches to tackling issues of learning and conceptual understanding should be sought and implemented. One of these approaches entails the use of interactive dynamic simulations in order to visualize relativistic phenomena that exist outside everyday experience. Encouraging group discussions among students should also help address the major conceptual difficulties experienced by students.

Tests and examinations should include more conceptual questions and fewer numerical or standard experimental applications which the students can memorize and reproduce.

References

Arriassecq, I., & Greca, I. M. (2012). A teaching–learning sequence for the special relativity theory at high school level historically and epistemologically contextualized. *Science & Education, 21*(6), 827–851.

Bencze, J. L., Bowen, G. M., & Alsop, S. (2006). Teachers' tendencies to promote student-led science projects: Associations with their views about science. *Science Education, 90*(3), 400–419.

Cataloglou, E., & Robinett, R. W. (2002). Testing the development of student conceptual and visualization understanding in quantum mechanics through the undergraduate career. *American Journal of Physics, 70*(3), 238–251.

Chhabra, M., & Bajeva, B. (2012). Exploring minds: Alternative conceptions in science. *Procedia – Social and Behavioral Sciences, 55*, 1069–1078.

Davisson, C. J., & Germer, L. H. (1928). Reflection of electrons by a crystal of nickel. *Proceedings of the National Academy of Sciences of the United States of America, 14*(4), 317–322.

Driver, R., Asoko, H., Leach, J., Mortimer, E., & Scott, P. (1994). Constructing scientific knowledge in the classroom. *Educational Researcher, 23*(7), 5–12.

Duit, R. (2009). *Bibliography – STCSE. Students' and teachers' conceptions and science education*. Retrieved 01 December 2015, from http://archiv.ipn.uni-kiel.de/stcse/

Fischler, H., & Lichtfeldt, M. (1992). Modern physics and students' conceptions. *International Journal of Science Education, 14*(2), 181–190.

Gleick, J. (1988). *Chaos: Making a new science*. Harmondsworth: Penguin.

Hewson, P. W. (1982). A case study of conceptual change in special relativity: The influence of prior knowledge in learning. *European Journal of Science Education, 4*(1), 61–78.

Johnston, I. D., Crawford, K., & Fletcher, P. R. (1998). Student difficulties in learning quantum mechanics. *International Journal of Science Education, 20*(4), 427–446.

McDermott, L. C., & Redish, E. F. (1999). Resource letter: PER-1: Physics education research. *American Journal of Physics, 67*(9), 755–767.

McKagan, S. B., Perkins, K. K., & Wieman, C. E. (2010). Design and validation of the quantum mechanics conceptual survey. *Physical Review Special Topics – Physics Education Research, 6*, 020121.

Muller, D. A., & Sharma, M. D. (2007). Tackling misconceptions in introductory physics using multimedia presentation. *Symposium presentations in UniServe Science Teaching and Learning*. http://sydney.edu.au/science/Uniserve_science

Müller, R., & Wiesner, H. (2002). Teaching quantum mechanics on an introductory level. *American Journal of Physics, 70*(3), 200–209.

Pitts, M., Venville, G., Blair, D., & Zadnik, M. (2014). An exploratory study to investigate the impact of an enrichment program on aspects of Einsteinian physics on year 6 students. *Research in Science Education, 44*(3), 363–388.

Posner, G. J., Strike, K. A., Hewson, P. W., & Gertzog, W. A. (1982). Accommodation of a scientific conception: Toward a theory of conceptual change. *Scientific Education, 66*(2), 211–227.

Scherr, R. E. (2007). Modeling student thinking: An example from special relativity. *American Journal of Physics, 75*(3), 272–280.

Scherr, R. E., Shaffer, P. S., & Vokos, S. (2001). Student understanding of time in special relativity: Simultaneity and reference frames. *American Journal of Physics, 69*(S24–S1), S35.

Shabajee, P., & Postlethwaite, K. (2000). What happened to modern physics? *School Science Review, 81*(297), 51–56.

Singh, C. (2001). Student understanding of quantum mechanics. *American Journal of Physics, 69*(8), 885–889.

Singh, C., Belloni, M., & Christian, W. (2006, August). Improving students understanding of quantum mechanics. *Physics Today*, pp. 43–49.

Taber, K. S. (2002). Compounding quanta: Probing the frontiers of student understanding of molecular orbitals. *Chemistry Education: Research and Practice in Europe, 3*(2), 159–173.

Tsaparlis, G. (1997). Atomic orbitals, molecular orbitals, and related concepts: Conceptual difficulties among chemistry students. *Research in Science Education, 27*, 271–287.

Villani, A., & Arruda, S. M. (1998). Special theory of relativity, conceptual change and history of science. *Science & Education, 7*(1), 85–100.

Zollman, D. A., Rebello, N. S., & Hogg, K. (2002). Quantum mechanics for everyone: Hands-on activities integrated with technology. *The American Journal of Physics, 70*(3), 252–259.

Chapter 12
Learning About Green Chemistry Independently: Students' Point of View

Charles M. Kgoetlana, Hlawulani F. Makhubele, Lemukani E. Manganyi, Emmanuel M. Marakalala, Shirley K. Sehlale, Derrick O. Sipoyo, Neani Tshilande, and Thembani S. Vukeya

1 Introduction

Green chemistry is an inventive science based on fundamental research towards the development of new sustainable chemical processes (Anastas and Warner 1998). The green chemistry's desire is to produce chemicals that are as useful as possible while being safer both for us to use and the environment (David S.T. Clair Black 2008). We (the students) encountered green chemistry through a project in a process technology course, which is taken in the second semester of the third year for chemistry majors. Green chemistry was not part of the syllabus, but the lecturer decided that getting information about green chemistry was important for a better preparation in the modern industrial processes.

The project was carried out independently, meaning that we had to find out all the information through literature search. Each one of us chose a principle. The 12 green chemistry principles are: Prevention, Atom Economy, Less Hazardous Chemical Synthesis, Designing Safer Chemicals, Safer Solvents and Auxiliaries, Design for Energy Efficiency, Use of Renewable Feedstocks, Reduce Derivatives, Catalysis, Design for Degradation, Real-time analysis for Pollution Prevention, Inherently Safer Chemistry for Accident Prevention (Anastas and Warner 1998). We had to search the information on our own, but since it was for the first time we worked on a research project, consultation was involved on how to prepare the slides for the presentation and on the relevance of the obtained information.

The literature search on each principle of green chemistry and its application involved the search for information about the importance of each principle and

Note: See editor's note at the end of the chapter.

C. M. Kgoetlana · H. F. Makhubele · L. E. Manganyi · E. M. Marakalala · S. K. Sehlale
D. O. Sipoyo · N. Tshilande (✉) · T. S. Vukeya
University of Venda, Thohoyandou, South Africa
e-mail: tneani11@gmail.com

L. Mammino, J. Apotheker (eds.), *Research in Chemistry Education*,
https://doi.org/10.1007/978-3-030-59882-2_12

about its applications. We were supposed also to search examples of applications that have already being done in the industries and have been successful. Each principle explains the importance of green chemistry in specific aspect, e.g. prevention explain the importance of preventing pollution and how to control or minimize pollution.

All the mentioned principles have their own different ways of dealing with the environmental health problems brought about by environmental pollution, but they are closely related and they outline effective ways to solve problems brought about by chemical industries (Clark 1999). For example, atom economy is closely related to prevention of waste. One should understand the principles well, so as to be able to develop new ways of proceeding in the chemical industry which are environmentally friendly and to understand what green chemistry means and what the concepts of green chemistry are. The next section describes how we went about researching independently and prepared the presentation and how we find it positive.

2 Overall Approach

Each student chose one principle from the 12 green chemistry principles (Anastas and Warner 1998), carried out a literature search about it and prepared the presentation. The search involved choosing relevant information and distinguishing it from the irrelevant one. For example, for principle 5, "Safer solvents and Auxiliaries", we can find in the literature the definition of green solvents (green solvents are environmentally friendly solvents) or of biosolvents, which are derived from the processing of agricultural crops (www.chm.bris.ac.uk, 2004). This was a relevant definition, because it explained what green solvents are. Not all the information can be used as definition. For example, we found a list of solvents used for chemical reactions, which is annotated with information about its health and safety profile, and the environmental problems associated with its use and disposal (http://molsync.com/demo/greensolvents.php); this is relevant information, but cannot be included in a definition.

The search for household and everyday life experiences was based on our own everyday life experience, including when we perform experiments in the lab and we saw that high quantities of solvents are used. Another aspect was the issue of littering around the schools and the streets.

The sources of information were very limited, as the internet was the only source of information where we accessed journals, articles and books about the concept of green chemistry, which was very helpful. Of all the information obtained, then we choose the information relevant to our individual principles. The discussion amongst ourselves also gave guidelines and clarity on how to conduct the research.

None of us knew how to prepare slides for a formal presentation, so consultation amongst ourselves and the lecturer on how to prepare slides was one of the tools that helped us to put information in a presentable way. Because each slide must be self-contained, we had to find a way of putting together information that had a meaning when put on one slide.

Presentation was also a problem, because none of us had presented a formal presentation before, so we had stage fright. The stage fright is actually the fear of having included irrelevant information so ending up presenting irrelevant information. The method of better presentation performance was also discussed amongst ourselves, as to which aspects are important to present and elaborate, as none of us had knowledge about research and presentation, but were trying to come up with a good way to put good presentation performance.

The poster in Fig. 12.1 presents the main features of the way in which the project was carried out, and its challenges and benefits, and provides some concrete examples from its realization.

3 Outcomes

On the research conducted, we had presentation on the selected topics from the 12 green chemistry principles (Anastas and Warner 1998). The presentation was performed in the presence of our process technology course coordinator and other lecturers who assessed us for marks, and also colleagues and some other students from other courses. Overall, the performance was good because none of us failed.

Not only the marks mattered, but the research left lot of knowledge with us about green chemistry that will remain for use in the near future. The research taught us the importance of green chemistry and how it should be applied in chemical industries. There are too many toxic chemicals used in the world and some are in the environment, so, many ways of solving problems should be employed, and that should come from us the chemists. The research taught us how to think green chemistry. so as to contribute to save the world, as certain ways of doing things in the chemical industries have already put it in danger.

Before the presentation we all had fear and also stage fright, but after the presentation the fear about formal presentation had now vanished, we gained confidence in ourselves and were willing to do another literature research. We learnt how to prepare formal oral presentation, how to work with PowerPoint. The confidence we gained was also from each other, when one was presenting, because if one of us can do it, also the others can do it.

The research made us believe that some things can be changed, so, things can improve; it made us see green chemistry to be a very useful tool to assess problems that chemistry brought and to avoid further problems. It made us see how dangerous chemicals are and that the job of a chemist is very important as it requires carefulness, not only in the lab on performing operations, but also towards the environment, and towards human health. We (the students) felt so positively about this experience that we wished to present it at an education conference, because we thought it will be useful also to other students to know how doing independent research on something will help you learn a lot. Our presentation was a poster and it is shown in Fig. 12.1.

Learning about green chemistry independently: students' point of view

This presentation concerns an activity carried out within the process technology course, at the University of Venda, is a third year course. Green chemistry is not part of the course syllabus, but is important for a better preparation in modern industrial process. We (the students) learnt about it through a project within the course. We were given the project in which one of us selected on of the twelve principles of green chemistry and carried out a literature research on that principle, to find information about its motivations and relevance, its applications in the chemical industry and , whenever relevant, also possible implications for everyday life. At the end of the course, each of us made an oral presentation on the selected principle.

Principle #1
Prevention

The project presented some challenges, because we had to learn how to distinguish the information relevant to the selected principle and the one that was not so relevant, and we also had to learn to organise the information in a way suitable for presentation. Another challenge was the stage fright as it was our first time to prepare and deliver a formal presentation.

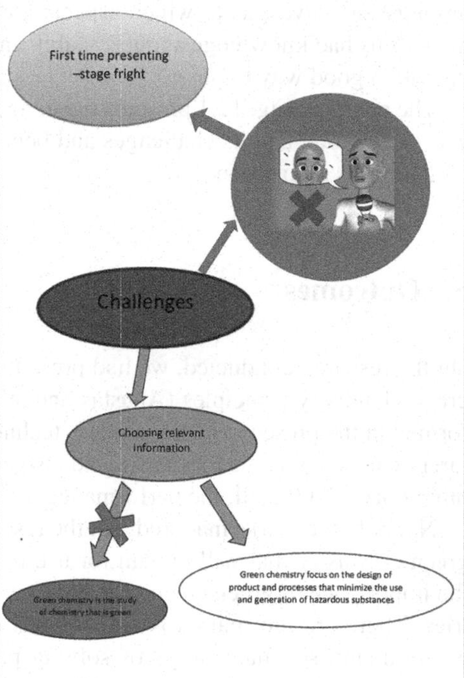

Working on the project gave us basic knowledge about the importance of green chemistry and its principles. Moreover, we displayed collaborative tendencies for which we were advising each other on how to go about it, and we aimed at performing at our best. The research was interesting not only for what we learnt about green chemistry, but also because we gained confidence in ourselves so as to be more effective in our presentations. All this positive aspects derived from the fact that we were doing our research independently. We think that the method of doing our research independently should be applied also in the future so that students can have experience on how to be independently researchers.

BEING INDEPENTENT CAN BE GOOD, FOR EFFECTIVE LEARNING..........

Vukeya TS, Kgoetlana CM, Marakalala ME, Sipoyo DO, Selahle SK, Makhubele HF, Tshilande N

Fig. 12.1 Poster presented at ACRICE2015

4 Discussion and Conclusion

As process technology course students in 2015, we found the concept of green chemistry most interesting and having potential in the way it deals with the problems of chemical industries. This includes the production of substances, the way they are used and the way they are being disposed at the end, and it involves the responsibility of the industries and of everybody who is using the substance. The 12 principles can be used to tackle pollution, prevent chemical pollution related diseases; and the fact that the green chemistry concepts are stated and explained one by one is good for one's understanding of the concepts. The literature search helped us understand green chemistry and to learn about preparing presentation. In general, it gave us confidence to do research independently.

This research project on green chemistry made us realize that the chemist's job is very important. This kind of project should be introduced to the secondary schools, as it does not only deal with industries, but is relevant to everyday lives; even at home chemicals are used and they need to be used in a correct way. The advantage of introducing this concept to schools will be that pollution by littering will be reduced. Not only will pupils' knowledge about green chemistry help them avoid littering, but they will grow up understanding the work of a chemist and they can also teach the community about saving resources and avoiding pollution.

In conclusion, green chemistry is a solution to most of the pollution problems in the world, and some knowledge about green chemistry should be shared with every citizen. Learning independently about green chemistry yields good results and we (the students) recommend it to everyone.

Acknowledgement We would like to thank ACRICE 2015 (2nd African conference on research in chemical education) for giving us the opportunity to present our poster at the conference. Our greatest gratitude goes to Prof Liliana Mammino, for guidance throughout the literature research process and the organization and preparation of the presentation and poster.

References

Anastas, P. T., & Warner, J. C. (1998). *Green chemistry: Theory and practice* (p. 30). New York: Oxford University Press.

Clark, J. H. (1999). Green chemistry: Challenges and opportunities. *Green Chemistry, 1*, 1. https://doi.org/10.1039/A807961G.

David ST Clair Black. (2008). *An introduction to green chemistry*. School of Chemistry, The University of New South Wales, UNSW Sydney 2052, Australia.

http://molsync.com/demo/greensolvents.php

http://www.chm.bris.ac.uk/webprojects2004/vickery/green_solvents.html

https://www.acs.org/content/acs/en/greenchemistry/what-is-green-chemistry.html

Editor's Note (Liliana Mammino)

This chapter is included in these proceedings to highlight the potentialities, for educational research, of sharing the reasons for the selected educational approaches with students, as well as the significance of the information that can be obtained by getting feedback from them in an articulate form, through a text having the main features of an article and resulting from reflection. This chapter is thus different from all the other chapters in these proceedings: it does not present educational research performed by a teacher, but an activity seen from the point of view of a group of students.

The project to which the chapter refers is described and discussed in «Mammino L. In V. Gomes Zuin and L. Mammino (Eds), *Worldwide Trends in Green Chemistry Education*, 2015, pp. 1–15, Royal Society of Chemistry». In summary: each student was given one of the green chemistry principles, and asked to find enough information to be able to prepare a presentation. The presentation contributed to the assessment for the semester. It was recommended that students focus mainly on the industrial significance and applications of the principles, because the course was a process technology course, i.e., a course related to the chemical industry. It was also recommended that, when the given principle has possible implications for everyday life, the students should try to identify, include and discuss them. Waste prevention and energy saving were among the principles with most direct possible applications to everyday life.

The students prepared the text of this chapter through collaboration among themselves. I (as the lecturer of this course) did not intervene, except by being available to answer questions whenever students wished to discuss something. My interventions in these discussions were mostly meant to stimulate reflection, not to provide answers. The text of this chapter (as presented here) has been left nearly as the students wrote it (with only minor editing when needed to make a meaning clear). Because of the documentation-importance of maintaining students' wording, the personal form in sentences has been retained (whereas all the other chapters use impersonal form). It may be added that, besides the challenges the students encountered in carrying out the project, writing this chapter also constituted a challenge. It was the first time they wrote something with the format of an article. The challenges related to language-mastery are huge for a variety of reasons – a major one among them being the fact that the context is second-language. These challenges are clearly evident from the wording of the text. On the other hand, the students managed to express what they wanted to express, and this is an important outcome. They wrote this text when they were close to completing their third year (undergraduate); now most of them are postgraduate students and would write in more sophisticated ways. But the text reported here constitutes the immediate feedback after the project.

The students' participation in the ACRICE-2 conference stemmed from my practice of informing students about what I do, for activities such as attending conferences or organising conferences, and explaining them why I do it. In the case of the ACRICE-2 conference, we talked about the roles of research in chemistry education, and I mentioned the importance of students' feedback. This led to the idea that they might participate in the conference. At that time, I invited them to choose the features of the activity, which they considered more important and to focus their poster on them. The students decided that the most important feature was the fact that they were learning something independently, that they were in charge both of finding and of learning the information. This aspect went directly into the title of the poster. In line with the idea of their independence, the students worked on their poster autonomously, and showed it to me for the first time when it was already completed; and it was practically ready, did not need any corrections (it also won the second prize for posters). Writing a chapter for the ACRICE-2 Proceedings was a natural continuation of their involvement. Since I personally believe that students should be involved in educational research also as active partners (with modalities to be explored), I encouraged the idea.

Chapter 13
The African Context: Investigating the Challenges and Designing for the Future

Liliana Mammino

1 Introduction

The previous chapters have outlined a variety of challenges facing chemistry education in African countries: poor laboratory facilities (Chaps. 1, 2, 3, and 6), inadequate content knowledge by many pre-university teachers (Chaps. 1, 3, 6, and 8), widespread rote learning and memorization without understanding by the learners (Chaps. 3, 5, 10, and 11), and others. The identification of challenges prompts enquiries about how to tackle them, and which approaches can be more prospective and effective and, at the same time, realistic (implementable) under the given conditions.

This conclusive chapter tries to outline a panoramic overview, focusing on those challenges that appear to have heavier impacts on learners' acquisition of knowledge, and outlining possible interventions. The overview is not meant to be a summary of the previous chapters, but rather a proposition of pertinent focuses for reflection, where the reflection is meant to constitute the foundation for the design of viable approaches. This overview cannot be exhaustive, either in terms of considered challenges or in terms of delineated approaches; but it attempts to propose options that are sufficiently promising and realistic to be treated as suitable objects of further exploration, in view of workable implementations.

Most of the challenges considered in the next section are not exclusive to the African context: they are present – with similar or varying characteristics – in many other contexts. Their discussion here refers mostly to the African context, but the terms of their scrutiny, as well as the outlined interventions, could be of interest also for other contexts.

L. Mammino (✉)
School of Mathematical and Natural Sciences, University of Venda,
Thohoyandou, South Africa
e-mail: sasdestria@yahoo.com

L. Mammino, J. Apotheker (eds.), *Research in Chemistry Education*,
https://doi.org/10.1007/978-3-030-59882-2_13

It is also necessary to recall that speaking of 'African context' is actually a not-completely or not-always-justifiable generalisation. Africa is a huge continent, with many diverse realities. The identified challenges may have different characteristics in different realities, and the design of addressing approaches needs careful adaptations to those characteristics. In view of this, the approaches proposed here have a sort of basic nature; they can be viewed as sort of canvases, into which details need to be filled according to the necessities of different realities.

The challenges themselves have different natures; for example:

- Challenges deriving from socio-economic reasons, such as overcrowded classes, or lack of laboratory facilities;
- Historically inherited challenges, such as the use of a language of instruction different from the learners' mother tongue;
- Challenges related to the teachers' subject knowledge, affecting prevalently (but not only) pre-university instruction;
- Challenges associated with the way in which the teaching and learning process is perceived and realised – the most serious of which being the diffuse practice of passive memorization (memorization without understanding) – practiced by learners, but often encouraged (or not discouraged) by teachers.

The author has long experience in chemistry-teaching in African countries: National University of Somalia (1974–1975), University of Zambia (1988–1992), National University of Lesotho (1993–1996) and University of Venda (UNIVEN) since September 1997. Most of the suggested approaches, as well as the arguments in their outlines, stem from her direct experiences.

2 Reflecting on Identified Challenges and Exploring Ways to Tackle Them

2.1 Recognition of Challenges and of Their Interrelations

This section considers some of the major challenges determining the success extent of the teaching and learning process. For the sake of clarity, different challenges are analysed in different subsections, whose titles express the nature of the given challenge. It has however to be taken into account that the challenges are not isolated (as if they were items assigned to sort of separated mental boxes); there may be different types of correlations among them, including cause-effect relationships or mutual enhancements. For instance, inadequate language-mastery is one of the major causes of passive memorization, because students memorise what they do not manage to understand on reading or listening; inadequate content knowledge by the teachers is also a major cause of passive memorization because, if the teacher cannot explain a certain concept in a sufficiently detailed and clear manner for students to understand it, students can only resort to passive memorization. Therefore, these three challenges – inadequate language-mastery, inadequate content knowledge by the teachers, and passive memorization – are related, as the first two are among the major causes of the third one. Figure 13.1 outlines the primary correlations among some of the major diagnosed challenges and the routes through which they impact on learners' acquisition of knowledge; an emotional factor – the learner's acquisition of interest in the subject – is included because of its important role in determining the efforts that the learner devotes to the subject, as well as his/her potential enjoyment in acquiring new knowledge about it.

It is not easy to find an optimal sequence for the consideration of the various challenges; different sequences may have comparable values; the sequence selected here is just one of the possible options. The outline for each challenge is necessarily concise, because of the nature of this chapter as an overview, and because of obvious space limitations.

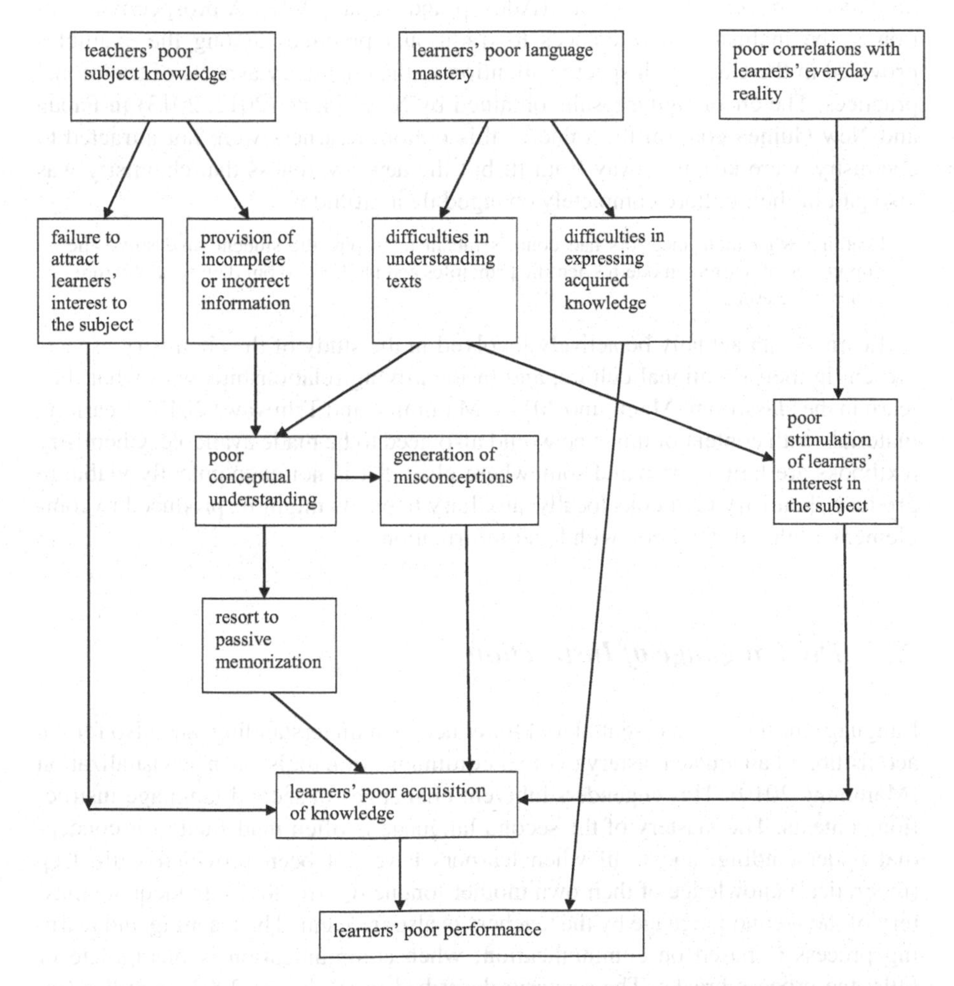

Fig. 13.1 Selected relevant correlations among diagnosed challenges and their impacts on learners' acquisition of chemical knowledge

2.2 The Need for Interplays Between 'Local' and 'Global'

Modern science is global: concepts, procedures and techniques are the same, whichever the country or the context. However, learners need to be able to relate them to their direct experience, in order to perceive those concepts as concrete. It is the great challenge of relating classroom chemistry to everyday life (Pinto Cañón 2003) and, possibly, also to history, within chemistry education, and within science education in general. The relationships are particularly important in the learners' early encounters with chemistry (or with another science).

This connection is largely missing in the African context, as clearly indicated by a number of studies (e.g., Abonyi et al. 2014; Adesoji and Akpan 1991; Gerdes 1994; Hyland 1994); its absence contributes "*to poor concept formation and attitude among beginners to science*" (Adesoji and Akpan 1991). A prospective solution is the inclusion of references to traditional practices among the examples provided in the class, with specific attention to the chemistry aspects of traditional practices. The encouraging results obtained by Marasinghe (2012, 2013) in Papua and New Guinea confirm the value of this option. Learners were not attracted to chemistry, were keeping away from it; but the new awareness that chemistry was also part of their culture completely changed their attitude:

> Learning what their ancestors had done is meant to inspire the students and stress the importance of learning modern scientific principles and methods to build on what the forefathers have done.

Learners can actually be actively involved in the study of the chemistry aspects present in their traditional culture, and in identifying relationships with what they learn in the classroom (Mammino 2012a; Mammino and Tshiwawa 2017). Learning materials with content of this type would also need to be made available. Chemistry textbooks are largely prepared somewhere else; if it is not economically viable to produce chemistry textbooks locally, auxiliary booklets might be produced to complement a 'global' textbook with local information.

2.3 The Language of Instruction

Language mastery is an essential tool for conceptual understanding, and also for the acquisition of adequate mastery of other communication tools, such as visualization (Mammino 2010). This engenders inherent challenges in second-language instruction contexts. The mastery of the second language is often inadequate for conceptual understanding, above all when learners have not been provided with deep (theoretical) knowledge of their own mother tongue (Qorro 2013). Inadequate mastery of the second language by the teachers is also frequent. The teaching and learning process is based on communication; when communication is incomplete or fails, the process breaks. The scenario described in (Rubanza 2002) is still a frequent sad reality:

> When teachers and learners cannot use language to make logical connections, to integrate and explain the relationships between isolated pieces of information, what is taught cannot be understood – and important concepts cannot be mastered.

This statement clearly recognises the minimum level of language-mastery sophistication needed to communicate science concepts: the ability to make logical connections, which entails the ability to build and understand complex sentences. This ability is acquired in the mother tongue as part of the natural process of learning it; it then needs to be turned into explicit awareness of the various logical connections, for these connections to be recognised and utilised in another language, expressing them through the tools pertaining to that language. It thus becomes necessary to recognise the role of the mother tongue as the primary and most natural communication tool, constituting a sort of 'mental home' (Seepe 2000), enabling maximum communication effectiveness, and enabling the acquisition of refined mastery of other languages.

The debate on the language of instruction is still on-going in some countries in Sub-Saharan Africa. On one side, there is clear recognition that only mother tongue instruction can ensure sufficiently diffuse and deep conceptual understanding (fundamental science knowledge) to enable African countries to get the number of adequately prepared specialists that is needed for development (e.g., Prah 1993, 1995, 2002; Owino 2002); and there is also the recognition that conceptual understanding is pre-requisite to adequate preparation. On another side, there is the conviction that science can only be expressed through English (a conviction pertaining to colonial inheritance and testifying lack of familiarity with other realities in other continents). There is also a diffuse perception viewing the learning of English as the main objective of all the courses in pre-university instruction, consequent to a perception linking the mastery of English to the possibility of job acquisition and social promotion. This side thus supports the idea that chemistry (or other subjects) should be taught through English because this would enable learners to acquire better mastery of English, and neglects the fact that, e.g., the primary objective of a chemistry course is for learners to learn chemistry.

The primary purpose of a science course is that learners learn that science. Therefore, the main objective is conceptual understanding. A viable and educationally promising option could separate the 'understanding' component of the learning process and the expression-of-acquired-knowledge component, and treat them differently (Mammino 2010): 'understanding' is achieved more effectively through the mother tongue; expressing the acquired knowledge can be done both through the mother tongue (which should never be neglected, as the aforesaid "mental home") and through English (or another foreign language). Once a concept is clearly understood, it is easy to express it correctly through more than one language. Trying to express it through two languages may also contribute to reinforce the level of understanding, because of the concept's analysis inherent in the use of different languages (the author speaks from direct experience, having working knowledge of four European languages, and having extensively interacted with students on how to express things correctly). On the other hand, the very nature of this approach requires good content knowledge from the teacher, and its exploration and possible implementation are hampered by inadequate teachers' preparation.

The frequently-encountered objection that African languages cannot express science because they do not have complete scientific terminology is not substantiated, and should not be allowed to have operational impacts in preventing the use of the mother tongue in science education. Terminology is not the information carrier in a discourse (Mammino 2006); it is an ensemble of names of individual items, and names can be formed at any moment, with the criteria that appear more suitable. Students are actually already using their mother tongues to explain things to each other, and have made their own terminology; since languages are 'alive', the way in which students are already using their own languages could provide valuable guidelines/references for the compilation of an official terminology.

2.4 The Teachers' Subject Knowledge

Most of the teachers who teach chemistry at pre-university level are not chemists. This is true in many contexts, not only in the African ones. It is likely unavoidable, because the number of students choosing chemistry for their tertiary studies is usually not high (chemistry is often considered difficult), and most of them opt for employment in the industry for their post-graduation careers.

Not being chemists unavoidably entails limitations in the teachers' conceptual knowledge. This, in turn, entails limitations in their pedagogical approaches, because a teacher who is not confident in his/her knowledge is often not willing to accept that learners may ask questions, or to use interactive teaching options. In general, learners cannot learn more than what their teachers know; it is also true that, in order to teach a certain content well, the teacher's knowledge should span a considerably broader scope than the content he/she is expected to teach. As a result, inadequate knowledge from the teacher's side engenders a downward loop towards increasingly poorer general knowledge, as the learners of a certain cohort of teachers will in turn become teachers with poorer knowledge than that of their former teachers, and the downward loop continues for subsequent cohorts of learners and teachers.

It becomes necessary to offer teachers a type of support that takes care of as many details as possible of the material that they are expected to teach. This needs to concern the entire material of a course – what becomes challenging for training or upgrading workshops, because their time limitations limits their scope. A possible option is that textbooks be prepared without assuming that the teacher already has adequate knowledge of the content. Thus, the textbook is meant for learners, but it takes care of explaining all the details, so that the teacher can learn from the textbook and clarify all his/her doubts before going to a class. This approach needs to be utilised for the entire textbook, so that it responds to the entire course content. A direct experience by the author (Mammino 1993) in Italy (where – as well – most secondary school chemistry teachers are not chemists) has proved that the option is viable (a number of teachers specifically expressed their appreciation for a textbook that explained to them while explaining to students).

The recognition of the problems engendered by the teachers' inadequate content knowledge also entails deep changes in the way new generations of teachers are trained. While it is undoubtedly important that future teachers learn about teaching approaches, it is also extremely important that they acquire sufficient knowledge of the subject content. Good knowledge of teaching methods not accompanied by good knowledge of the subject content is bound to remain sterile. An analogy in mathematical terms might help emphasize the truth of this statement. The teaching methods might be compared to operations to be performed on a set, where the set is the content knowledge; any operation performed on an empty set will give an empty set as result. The content knowledge is also pre-requisite to the implementation of the best educational approaches: the greater the teacher's content knowledge, the more he/she can be confident to implement effective approaches like interactive teaching and active learners' involvement.

2.5 *The Lack of Laboratory Facilities*

This problem is mostly related to financial constraints. The situation is different in different countries, and often also in different areas of the same country. Teachers' responses are also different. Some teachers are discouraged by the lack of facilities and do not attempt to design anything to compensate for it. Other teachers utilise materials available in the environment, or in normal shops, to set up illustrative demonstrations, often with a good deal of creativity.

It is important to provide adequate opportunities for the latter teachers to share their experiences and creativity, so as to encourage the others. One of the main objectives of laboratory demonstrations/practicals at pre-university level is to ensure that learners learn to make observations and to describe them, and – when suitable – also to propose inferences and interpretations. These objectives can be attained also with simple demonstrations using materials available in the surroundings (for instance, the juice of a number of berries can be used as acid-base indicator). The teacher's guidance is the most important resource for the knowledge-generation efficacy of these approaches.

2.6 *Tailoring Educational Approaches to the Learners' Needs*

It is fundamental to tailor educational approaches to the characteristics and needs of the specific group of learners that one is teaching (to the 'real learners', as opposed to a generalised 'learners' concept, which tends to institute a sort of 'ideal' category devoid of contextual characteristics). This requires continuous verification of the learners' responses to the approaches used by the teachers. Action research (Levin

1947) (described in some detail in Chap. 10) is an optimal instrument to this purpose. It would be important to encourage teachers, at any level of instruction, to use action research to improve their educational approaches specifically for the group of learners they are teaching in a given period, and also to share their results, possibly through publications or through specifically-devoted online sites, because this would be useful to other teachers facing comparable challenges.

Interactive teaching is the general approach which enables real-time recognition of the learners' responses, and immediate adaptations by the teacher (i.e., immediate teacher's responses to the diagnoses enabled by the learners' responses). Ideally, the entire time of a lesson, or of a lecture, should entail a sort of tuning, expressed through responses and responses-to-responses. The teaching and learning process becomes a shared process between teacher and learners. The use of interactive teaching requires that the teacher is confident with his/her subject knowledge, and that he/she is available to design *ex-promptu* and even unconventional options to stimulate learners to participate in the interactions (Mammino 2011). When a group of learners (for whichever reasons) finds it difficult to respond verbally, the resort to in-class written questions and answers proves a functioning substitute (Beall 1991; Mammino 2013), to ensure a form of interaction which, although different, maintain the main essential values.

Perceptions that may restrain students' availability to interactions also need addressing. These include perceptions that might freeze the roles of teacher and learners into rigid stereotypes, perpetuating passive attitudes from the learners' side. Unconventional options may be the most effective for addressing them. For instance, in order to counteract the diffuse expectation that the teacher/lecturer knows everything by definition, the author has established the practice of giving two marks bonus to the first student who finds an error (not a typo, but any other type of errors) in a question proposed in a text or in an exam. This information is always provided to students a couple of days before the first test of a semester. When some students express surprise, perceiving their signalling of an error in a text prepared by the lecturer as something that goes against the expected teacher's and learners' roles, the lecturer's response specifies the reality (the true nature) of these roles, in terms like "My objective is that you learn the concepts; if you find an error in what I have written, it means that you understand the concepts well enough to be able to identify errors; in other words, it means that I have done my job properly". Although errors actually occur rarely (mostly in terms of a missing value in a long list of values from literature), informing about the lecturer's practice is functional to stress the learner-centred role of the course activity.

A further step would entail sharing with the learners the motivations for the selection of the teacher's approaches. The author comes from such an experience, as her secondary school chemistry teacher used to share his motivations for his pedagogical approaches, thus stimulating interest in chemistry education at the same time as he was stimulating interest in chemistry. It is important to stimulate interest in chemistry education as a general attitude, because of the importance of chemistry literacy for informed citizens and for sustainable development. Chemistry students and young chemists with sufficient education abilities can play crucial roles to promote sustainability in their communities. At suitable instruction levels, learners may also be invited to analyse and share their experience within a given approach or project; this provides valuable information to the teacher at the same time as it enhances students' active role in the learning endeavour. Chapter 12 is a contribution of this type.

2.7 *Integrating Chemistry Teaching and Language Teaching*

As mentioned earlier, conceptual understanding in chemistry (and in the sciences in general) is conditioned by the learners' (and teachers') language mastery level. Language mastery deterioration in the young generation is recently being noticed in many contexts, and seriously affects science learning. Because of this, it becomes important to integrate language teaching into the teaching of science subjects. The integration can be realised in different ways, according to the level of instruction and to the context.

The author has practiced this option extensively at the tertiary level, to train students to identify correct or incorrect expressions of concepts by analysing the meaning conveyed by a given wording and comparing it with a fitting reference. The option is suitable for many purposes. It is useful to address misconceptions inherited from past instruction, to warn against incorrect statements found in whichever source, and to try to accustom students to critically proofread what they have written in their answers or other works.

Diffuse tendency to oversimplification in previous instruction burdens students with a variety of inherited incorrect definitions. These are not misconceptions deriving from inadequate or incorrect understanding; they are misconceptions that have been directly taught. Therefore, the analysis of the actual meaning of the memorised wording, and the comparison of this meaning with a suitable reference, becomes the only option to counteract the errors. For instance, for the inherited incorrect definitions of direct and inverse proportionality ("two quantities x and y are directly proportional if y increases as x increases", and "two quantities x and y are inversely proportional if y decreases as x increases"), the selected references are two sets of diagrams, with each set containing several different diagrams compatible with one of the incorrect definitions (Mammino 2014); this shows that the inherited definitions do not enable univocal identification of the two proportionality relationships; then, students are prompted to build correct definitions.

When students acquire familiarity with this type of analysis, they can easily identify the problem in statements proposed to them. In the most successful cases (i.e., with the groups that best familiarise with this approach), students learn to ask the lecturer when they find statements in outside sources, of whose correctness they are not sure; then, these statements are discussed in the group. For instance, someone found a statement defining chemistry as the science of compounds; since it differed from the definition given in the class (chemistry is the science of substances), he mentioned it in the class; students were invited to reflect on it, and some of them quickly recalled that there are whole industries dealing with elements (iron, copper, oxygen, chlorine, etc.); this also provided the opportunity to revise and stress the differences and relationships of the 'substance', 'element' and 'compound' concepts. The analysis of the actual meaning of the way in which statements are worded can also be applied to the errors that students make in their answers, and proves a powerful tool to clarify the chemistry concepts concerned (Mammino 2015). Fostering the habit to critically proofread their own writings appears to be more challenging, even for works like the lab reports, which do not involve the tension usually present during tests and exams; it becomes important to endeavour to design new options capable of promoting the acquisition of this practice.

In a number of contexts, part of the students entering university does not have adequate background knowledge of one or more science subjects. They are usually asked to take a bridging course to acquire the foundations of the sciences in which they are lacking. Practically all those students also lack adequate language mastery. The bridging course would be an ideal opportunity to integrate chemistry teaching and language teaching (Mammino 2012b), with the double benefit that language analysis clarifies the chemistry concepts, and that students learn how to express chemistry correctly. In second-language contexts, the participation of an expert in the learners' mother tongue would greatly contribute to enhance the benefits.

The integration of language analysis into chemistry teaching would actually be beneficial at all levels of instruction. The practical difficulty is that, in most cases, chemistry teachers are not familiar with language analysis (with the identification of the actual meaning conveyed by a specific wording and its comparison with the intended chemistry concept) and, on the other hand, language teachers would not be able to identify chemistry errors conveyed by incorrect wordings. An additional difficulty resides in the novelty of the proposition, and in the scarce familiarity with the causes of students' learning difficulties by those who would have the responsibility to approve projects of this type. Therefore, despite the fact that the integration of language teaching and chemistry teaching has all the potentialities to greatly enhance the quality and effectiveness of chemistry learning, its extensive implementation still faces many challenges. Pilot projects conducted by chemistry lecturers familiar with language analysis can be functional to corroborate the effectiveness of the approach, thus enabling successive broadened explorations. Training chemistry teachers in this type of analysis may prove quite demanding; on the other hand, it would result in simultaneous substantial upgrading of their knowledge and understanding of chemistry concepts, and of the effectiveness of their teaching, since expression rigour is by itself a pedagogical tool (Mammino 2000).

2.8 *Modern Chemistry for the Needs of Modern Times*

Chemistry plays fundamental roles for development. In current times, it plays fundamental roles in the efforts to make development sustainable. The young generation needs to get adequate exposure to these roles, as well as adequate exposure to the latest developments in our understanding of substances. This entails re-examination of the syllabi. In addition, the particulars of the syllabi need to correspond to the most recent conceptions; this entails re-examination of the details that are provided to learners and students in their familiarization with the concepts of basic chemistry.

As far as syllabi are concerned, the incorporation of two major recent themes is of paramount importance: green chemistry and the study of molecules. Green chemistry is the chemists' response to the requirements of sustainability, and gives chemistry a vital role to make development sustainable. Information about green chemistry can be provided since the learners' first encounter with chemistry. The 12 principles of green chemistry (Anastas and Williamson 1996) can be introduced and explained individually in terms adapted to the level of instruction. The study of molecules can be introduced in qualitative terms, ascribing dominant role to visualization (Mammino 1993) and selecting the images in such a way that they can be understood easily and, at the same time, can convey the awareness that modern chemistry relies increasingly more extensively on the study of molecules.

The way in which the concepts of basic chemistry are presented needs to be compatible with modern chemistry. This may require some 'clean-up', to remove images, classifications or other details that do no more respond to the body of knowledge of modern chemistry. It is a type of clean-up conceptually parallel to the criterion with which Lavoisier dismissed the phlogiston theory: it was not needed to explain observed phenomena. There is unavoidable inertia (like in many human things), which somewhat hampers the recognition of what is obsolete or simply unnecessary. Teachers tend to perpetuate images, concepts and details that they have learnt before becoming teachers, and do not realise the lack-of-use of some of them, nor the damaging potentialities of others. For instance, the images representing covalent bonds in terms of circular orbits of the bonded atoms coming into contact, with two electrons fixed in the contact point, generate incorrect perceptions, from the inference that molecules are flat (a logically correct inference from the incorrect image, resulting in a scientifically incorrect concept), to the idea that electrons have fixed positions in those circles. The author finished secondary school more than 50 years ago; in those times, she was lucky enough that her school had been selected for a pilot project aimed at teaching chemistry in modern terms; thus, as a pupil or student, she has never encountered those images, or earlier classifications of chemical reactions, or other details that pertain to the chemistry up to the late 1950s; it becomes surprising that details and images that were considered obsolete within a "modern" chemistry course in the mid and late 1960s are still encountered rather frequently, having lingered by inertia. Removing concepts and details that are misleading, or simply not needed in modern chemistry, would help make the conceptual content more to-the-point, thus facilitating understanding.

3 Discussion and Conclusions

The previous sections have considered some crucial challenges and offered quick outlines of selected promising approaches to tackle them. These are not the only challenges present, and the outlined approaches are not the only possible ones. A lot of exploration is needed to design the best modes for implementing the outlined suggestions, and a lot of exploration is needed to identify additional possibilities.

The design of educational approaches is a multi-side task, because of the variety of aspects to be taken simultaneously into account. The importance of chemistry for development in general, and for sustainable development in particular, entails the importance of effective chemistry education, both in general terms (to prepare informed citizens capable of making the best decisions for sustainable development) and for the training of adequately prepared specialists. Sharing educational experiences is the key pathway to extensive improvement of the quality and effectiveness of chemistry education in Africa and beyond.

References

Abonyi, O. S., Achimugu Njoku, L., & Ijok Adibe, M. (2014). Innovations in science and technology education: A case for ethnoscience based science classrooms. *International Journal of Scientific and Engineering Research, 5*(1), 52–56.

Adesoji, F. A., & Akpan, B. B. (1991). Teaching science mathematics and technology in mother tongue: A path to curriculum evaluation. *32nd annual conference proceedings of STAN*, pp. 70–73.

Anastas, P. T., & Williamson, T. (1996). In P. T. Anastas & T. Williamson (Eds.), *Green chemistry*. Washington, DC: American Chemical Society.

Beall, H. (1991). In-class writing in general chemistry: A tool for increasing comprehension and communication. *Journal of Chemical Education, 68*(1), 148–149.

Gerdes, P. (Ed.). (1994). *Explorations in ethnomathematics and ethnoscience in Mozambique*. Maputo: Instituto Superior Pedagogico.

Hyland, R. E. (1994). A review of Paulus Gerdes, editor. In *Explorations in ethnomathematics and ethnoscience in Mozambique*. Maputo: Instituto Superior Pedagogico. Available at https://web.ccsu.edu/afstudy/upd2-2.html.

Levin, R. (1947). Frontiers in group dynamics. II. Channels of group life: Social planning and action research. *Human Relations, 1*, 143–159.

Mammino, L. (1993). *Chimica Viva*. Florence: G. D'Anna.

Mammino, L. (2000). Rigour as a pedagogical tool. In S. Seepe & D. Dowling (Eds.), *The language of science* (pp. 52–71). Johannesburg: Vyvlia Publishers.

Mammino, L. (2006). *Terminology in science and technology*. Thohoyandou: Ditlou.

Mammino, L. (2010). The mother tongue as a fundamental key to the mastering of chemistry language. In C. F. Lovitt & P. Kelter (Eds.), *Chemistry as a second language: Chemical education in a globalized society* (pp. 7–42). Washington, DC: American Chemical Society.

Mammino, L. (2011). Teaching chemistry in a historically disadvantaged context: Experiences, challenges, and inferences. *Journal of Chemical Education, 88*(11), 1451–1453.

Mammino, L. (2012a). Chemical heritage in sub-Saharan Africa and its significance for chemical education. *Southern African Journal for Folklore Studies, 33*(2), 147–158.

Mammino, L. (2012b). Focused language training as a major key for bridging the gap between secondary and tertiary instruction. In D. Mogari, A. Mji, & U. I. Ogbonnaya (Eds.), *ISTE international conference proceedings* (pp. 278–290). Pretoria: UNISA Press.

Mammino, L. (2013). Teacher-students interactions: The roles of in-class written questions. In M.-H. Chiu (Ed.), *Chemistry education and sustainability in the global age* (pp. 35–48). Dordrecht: Springer.

Mammino, L. (2014). The interplay between language and visualization: The role of the teacher. In B. Eilam & J. Gilbert (Eds.), *Science teachers' use of visual representations* (pp. 195–225). Dordrecht: Springer.

Mammino, L. (2015). Clarifying chemistry concepts through language analysis. In J. Lundell, M. Aksela, & S. Tolppanen (Eds.), *LUMAT Special Issue of ECRICE, 3*(4), 482–500.

Mammino, L., & Tshiwawa, T. (2017). Chemistry practices in the Vhavenda indigenous society. *Indilinga, African Journal of Indigenous Knowledge Systems, 16*, 221–237.

Marasinghe, B. (2012, April 16–21). *Changing attitudes towards learning chemistry among school children and undergraduates in Papua and New Guinea*. Paper presented at the 12th Eurasia conference on chemical sciences, Corfu, Greece.

Marasinghe, B. (2013). *Ethnochemistry and ethnomedicine of Ancient Papua New Guineans and their use in motivating secondary school children and university undergraduates in PNG*. Available at http://stemstates.org/assets/files/Full%20PaperBasilMarasinghe.pdf. Accessed 02 June 2015.

Owino, F. R. (Ed.). (2002). *Speaking African – African languages for education and development*. Cape Town: CASAS.

Pinto Cañón, G. (Ed.). (2003). *Didáctica de la Química y Vida Cotidiana* [Chemical education and everyday life]. Madrid: Universidad Politécnica de Madrid, Escuela Técnica Superior de Ingenieros Industriales, Sección de Publicaciones.

Prah, K. K. (1993). *Mother tongue for scientific and technological development in Africa*. Bonn: German Foundation for International Development.

Prah, K. K. (1995). *African languages for the mass education of Africans*. Bonn: German Foundation for International Development.

Prah, K. K. (2002). In K. K. Prah (Ed.), *Rehabilitating African languages* (pp. 1–6). Cape Town: CASAS.

Qorro, M. (2013). Language planning and policy in Tanzania: When practice does not make perfect. In Z. Desai, M. Qorro, & B. Brock-Utne (Eds.), *The role of language in teaching and learning science and mathematics* (pp. 151–170). Somerset West: African Minds.

Rubanza, Y. I. (2002). In K. K. Prah (Ed.), *Rehabilitating African languages* (pp. 39–51). Cape Town: CASAS.

Seepe, S. (2000). In S. Seepe & D. Dowling (Eds.), *The language of science* (pp. 40–51). Johannesburg: Vyvlia Publishers.

Printed in the United States
by Baker & Taylor Publisher Services

Printed in the United States
by Baker & Taylor Publisher Services